Responding to Crises in the Modern Infrastructure

# Responding to Crises in the Modern Infrastructure

## Policy Lessons from Y2K

Kevin F. Quigley

*Assistant Professor, School of Public Administration,*
*Dalhousie University, Canada*

First published 2008 by
PALGRAVE MACMILLAN

Palgrave Macmillan in the UK is an imprint of Macmillan Publishers Limited,
registered in England, company number 785998, of Houndmills, Basingstoke,
Hampshire RG21 6XS.

Palgrave Macmillan in the US is a division of St Martin's Press LLC,
175 Fifth Avenue, New York, NY 10010.

Palgrave Macmillan is the global academic imprint of the above companies
and has companies and representatives throughout the world.

Palgrave® and Macmillan® are registered trademarks in the United States,
the United Kingdom, Europe and other countries.

ISBN-13: 978-0-230-53587-9 hardback

This book is printed on paper suitable for recycling and made from fully
managed and sustained forest sources. Logging, pulping and manufacturing
processes are expected to conform to the environmental regulations of the
country of origin.

A catalogue record for this book is available from the British Library.

Library of Congress Cataloging-in-Publication Data
Quigley, Kevin F.
    Responding to Crises in the Modern Infrastructure : Policy Lessons from Y2K /
    Kevin F. Quigley.
        p.   cm.
    Includes bibliographical references and index.
    ISBN 978-0-230-53587-9 (alk. paper)
    1. Risk management.   2. Year 2000 date conversion (Computer systems)
    3. Infrastructure (Economics)   4. Crisis management.   I. Title.
    HD61.Q54 2008
    363—dc22                                                    2008020598

10   9   8   7   6   5   4   3   2   1
17  16  15  14  13  12  11  10  09  08

Transferred to Digital Printing 2012

*For Sara, Eva and Rachel*

# Contents

# List of Tables

# List of Figures

# List of Acronyms

| | |
|---|---|
| ATC | Air Traffic Control |
| BDD | Business Development DISC |
| BLS | Bureau of Labor Statistics |
| CAA | Civil Aviation Authority |
| CCTA | Central Communications Technology Agency |
| CD | Compact Disc |
| CIA | Central Intelligence Agency |
| CIO | Chief Information Officer |
| CITU | Communications and Information Technology Unit |
| Cm | Command Paper |
| CO | Cabinet Office |
| COTS | Commercial Off-the-Shelf |
| DETR | Department of the Environment, Transport and the Regions |
| DOL | Department of Labor |
| DOT | Department of Transportation |
| EDS | Electronic Data Services |
| EO | Executive Order |
| EOP | Executive Office of the President |
| FAA | Federal Aviation Administration |
| FT | *The Financial Times* |
| GAO | General Accounting Office (now, the Government Accountability Office) |
| GDP | Gross Domestic Product |
| GMIT | Government Management, Information and Technology Sub-Committee |
| GMT | Greenwich Mean Time |
| HM Treasury | Her Majesty's Treasury |
| HRO | High Reliability Organizations |
| HSE | Health and Safety Executive |
| IATA | International Aviation Transport Authority |
| IDC | International Data Corporation |
| INT | Interview |
| IT | Information Technology |
| IVV | Internal Verification and Validation Process |
| JAA | Joint Aviation Authority |

| | |
|---|---|
| LOBs | Lines of Business |
| LT | *The Times* of London |
| MB | Micro Business (one to nine employees) |
| MCU | Media Communications Unit |
| MFH | Market Failure Hypothesis |
| NAO | National Audit Office |
| NATS | National Air Traffic Service |
| NHS | National Health Service |
| NIF | National Infrastructure Forum |
| NMD | 'No Material Disruption' to the essential public services upon which the public relies (the UK Y2K strategy) |
| NYT | *The New York Times* |
| OECD | Organization for Economic Cooperation and Development |
| OFTEL | Office of Telecommunications |
| OIG | Office of the Inspector General |
| OMB | Office of Management and Budget |
| ONS | Office of National Statistics |
| ORH | Opinion-Responsive Hypothesis |
| PAC | Public Accounts Committee |
| PC | Personal Computer |
| PRA | Probability Risk Assessments |
| SAG | Senior Advisors Group |
| SAIC | Science Applications International Corporation |
| SCST | Standing Committee on Science and Technology |
| SEC | Securities Exchange Commission |
| SME | Small and Medium-sized Enterprises |
| UK | United Kingdom |
| UN | United Nations |
| US | United States |
| USAT | *USA Today* |
| WAN | Wide Area Networks |
| WG | US Government Y2K Working Groups |
| WSJ | *The Wall Street Journal* |
| Y2K | Year 2000 Computer Bug |

# Acknowledgements

Much of the work in this book was completed as part of my PhD research at the Institute of Governance, Public Policy and Social Research at Queen's University Belfast (QUB). I was very fortunate to be one of the inaugural PhD students at the Institute. It was a lively and intellectually engaging place.

There are many people to thank for their help, and I hope I have conveyed my gratitude to them personally as I worked towards the completion of this project. I will name a few key contributors here. I wish to thank my PhD supervisors at QUB, Professor Rick Wilford from the School of Politics, International Studies and Philosophy and Professor George Philip from the School of Management and Economics. I would also like to thank the Director of the Institute, Professor Elizabeth Meehan. Professor Mel Dubnick from the University of New Hampshire was a visiting scholar at the Institute for two years while I was there. I consulted with Mel informally and frequently. Each of these four people played a vital role in helping me to conduct this research and bring it to completion in my PhD dissertation.

Following the completion of my PhD, I took up a postdoctoral fellowship at the University of Edinburgh. My mentors there, Professor Charles Raab from the School of Social and Political Studies and Professor Robin Williams, Director of the Centre for Social Sciences, provided thoughtful advice as I began to update the work and adapt its style to one that is more appropriate for a book.

I would also like to acknowledge the 68 interview subjects, who gladly gave their time to provide informed and thoughtful views and whose contributions were essential to the outcome.

Finally, I also owe a debt of gratitude to my family—my wife, Sara, my mom and dad and my two brothers, Jim and John—whose support started long before this research and was unflinching throughout. I dedicate this book to Sara, whose full support was essential for me personally and who offered it in a variety of ways, unconditionally, and to our daughters, Eva and Rachel.

# 1
# Introduction

Things did not go right by accident
—Y2K Press Release from the Cabinet
Office (Beckett, 2000a)

'The computer has been a blessing', wrote Senator Pat Moynihan in a July 1996 letter to President Clinton, 'if we don't act quickly, however, it could become the curse of the age' (Moynihan, 1996). Moynihan, the senior Senator from New York and respected academic in his own right, was commenting on the results of a Congressional study into a date-generated computer bug that became known as Y2K (Year 2000). President Clinton would eventually describe it as 'one of the most complex management challenges in history' (1998). Margaret Beckett, Chair of the British Cabinet Committee on Y2K, would refer to the UK government's response to it as 'the largest co-ordinated project since the Second World War' (Hansard, 1999).

Y2K, or the 'millennium bug', referred to the fact that in computer programmes created in the 1950s and onwards most year entries had been programmed in two-digit shorthand – 1965, for instance, was entered as '65'. Initially this shorthand was adopted to save on expensive computer memory. Through time it simply became standard practice. As the year 2000 approached, however, anxiety grew that systems would be unable to distinguish between twentieth century entries and twenty-first century entries (for example, would '01' be treated as '1901' or '2001'?). Such ambiguity, it was feared, would result in systems failing or producing inaccurate or unreliable information. The perceived dependence on technology and interdependence of the national and international infrastructures across sectors led decision-makers in both countries to conclude that Y2K was a challenge without precedent (see, for

1

example, HC Science and Technology Committee Report, 1997/1998; GAO, 1997a; 1997b). The US and the UK governments spent $10 billion on central government operations alone and (at a minimum) three years on preparations. And, in the end, virtually nothing happened. Did this mean success? Despite the scope and cost of Y2K it has received almost no critical analysis, academic or otherwise, post 1 January 2000. With renewed emphasis on risk management and infrastructure protection post September 11 and a call to use Y2K planning as a blueprint for such tasks,[1] the two governments' approaches to Y2K require further consideration.

## Y2K and the current debate about Critical Infrastructure Protection

Y2K may seem like a rather mundane dress rehearsal in a post-9/11 environment. In fact, the Y2K case provides exceptional insight into the constraints and opportunities that governments face when seeking to ensure the resilience of critical infrastructure. Data in this area is scarce. Y2K is perhaps the *only known example* of an (arguably) successful economy-wide CIP initiative that exists in recent history.

Critical Infrastructure Protection (CIP)—activities that enhance the physical *and* cyber-security of key public and private assets—is the focus of urgent attention among Western governments in the light of recent power failures, computer viruses, natural disasters, epidemics and terrorist attacks, both threatened and realized. Government studies and popular analyses note the complex, interdependent and fragile make-up of these infrastructures and the technologies that underpin them. Consider the 2003 North American power outage: overgrown trees in Ohio helped trigger a power failure that affected 50 million people and cost the US economy anywhere from $4 to 10 billion.[2]

Can government reaction to Y2K help us to understand better how to manage this potentially fragile infrastructure? In other words, can we draw meaningful lessons from Y2K? Certainly there are aspects of the event that are unique: the specific date-related nature of the problem, for instance. There is reason to believe there is common ground, also. Many of the critical features that prompted the initial concern about Y2K are similar to conditions we face today. In many respects Y2K merely punctuates a trend that has existed since the emergence of decentralized technologies, globalization and government outsourcing and privatizations, which have brought about simultaneously greater segmentation and interdependency in economic and social systems. In these respects, Y2K could be seen not as an isolated event but rather as an indication of the ongoing challenges of maintaining a stable infrastructure in such a complex and interdependent setting.

The problems are not merely technical. Many social, organizational and jurisdictional obstacles which pre-date the turn of the millennium prevent successful CIP. For most Western countries critical infrastructure is owned and operated by a large number of organizations from both the public *and* the private sectors. Corporate executives and their shareholders are reluctant to spend on CIP because its benefits are often indeterminate. They are also reluctant to disclose the vulnerabilities of their assets because of the risk to their organization's security, liability, reputation and share value. There is also a problem with trust. Industry executives worry that sensitive information shared with government may be used (surreptitiously) for reasons other than CIP. Government officials are equally reluctant to share sensitive information. Reporting lines in bureaucracies are bottom-up; outward accountability is not their strong-card. Also, leaked intelligence can bring about human devastation on a massive scale. Finally, overlapping responsibilities between different organizational units and levels of government can obscure accountability and complicate planning. In short, despite its acknowledged importance, CIP is an area in which it is difficult to achieve meaningful cooperation and transparency.

Government efforts to change this dynamic have produced mixed results. The US government's attempts to facilitate information exchange on CIP have been described as unsuccessful. The government has been slow to identify critical assets, map interdependencies, and identify and manage vulnerabilities. The US ISACs (Information-sharing and Advice Centers)—fora whose membership includes the private sector alone—have had uneven success in achieving desired participation rates, obtaining reliable data and developing a trusted mechanism by which to exchange reliable and sensitive information (GAO, 2003). The UK experience on the surface shows more promise. The UK's Civil Contingencies Secretariat and its related National Steering Committee on Warning & Informing the Public (NSCWIP) is based at Cabinet Office and therefore is closer to the centre of government than the American model. Unlike the US case, UK government aims to work more closely and collaboratively with industry (Cabinet Office, 2007).

While things look perhaps more promising for the Westminster system, legislative scrutiny of these arrangements varies. The US Government Accountability Office has been more active. Its reports on CIP can be counted in the dozens. Among them have been several reports explicitly on sharing potentially sensitive information across organizations about the critical infrastructure (GAO, 2003; 2004a). The United Kingdom's National Audit Office, in contrast, has published nothing on joint government/private sector CIP initiatives.

Many of the problems cited above existed in the run-up to Y2K. This book seeks to examine these and other contextual pressures that influenced each government's response to the challenges that emerged as well as the tensions and trade-offs in the solutions each government adopted. The chapters that follow will examine the roles of the law and insurance, the media, public opinion and organized interests, and consider the extent to which each of these helped shape government reaction to Y2K.

I hope to make a modest contribution to a debate that is far from mature. Indeed, academic scrutiny in this area has been uneven. To start, the study of security often examines the role of the state-level defence, which often excludes or marginalizes domestic security, which is of increasing importance post 9/11. Moreover, the term CIP, in particular, originates from the field of information technology. This literature has a bias towards quantitative methods and formal risk modelling, which can no longer be considered a universally accepted method of understanding risk, as will be discussed further in Chapter 2. There are some notable exceptions in the literature in which the search for domestic security is seen as a dynamic process in a complex multi-organizational setting (Aviram, 2005; Aviram and Tor, 2004; Auerswald, 2006; Comfort, 2002; de Bruijne and van Eten, 2007; Egan, 2007). This trend is relatively new, however. On the balance, the field has been slow to integrate the political science literature and sociology literature that consider the matter of power and interests and how they potentially influence the management of critical infrastructure. Finally, despite declarations such as that found in European Union Framework 7, which identifies CIP as an area in which greater international research and collaboration is required, CIP research to date has been overwhelmingly focused on the US case alone, with a specific emphasis on the implications of deliberate acts of terror.

## The approach and the central research question

This book is primarily about Y2K. Because so little analysis has been done on the Y2K case thus far I will bring an inductive approach to the case and, as far as possible, will let the data (both the historical documents and the opinions of the interviewees) generate the findings.

For this case I have elected to use Hood, Rothstein and Baldwin's (2001) meso-level Risk Regulation Regime framework to examine the US and the UK Governments' management of Y2K.[3] In so doing I will attempt to answer the central research question for this book, which is borrowed and adapted from the Hood *et al.* (2001) framework: *Does regime context shape regime management?* The Hood *et al.* (2001) framework is sufficiently flexible in that it casts a wide net for an inductive approach: the framework considers the law, the market, the media, public opinion, interests and institutions when examining factors that are potentially critical to understanding governments' approaches to risk management. This framework is well suited for this research because neither the context surrounding Y2K nor its management has been researched.[4] Moreover, the Hood *et al.* framework is a comparative tool.

In their recent study of risk regulation in the United Kingdom, Hood *et al.* deploy the concept of 'regimes' to explore variety in the different policy areas[5] (Hood *et al.*, 2001, 5). Using this diverse literature as their springboard, they define regimes thus: 'the complex of institutional geography, rules, practice and animating ideas that are associated with the regulation of a particular risk or hazard' (2001, 9). This broad definition allows for flexibility as Hood *et al.* read across various policy contexts while drawing together a variety of institutional perspectives in order to understand what shapes risk regulation.

Hood *et al.* hypothesize that within these regimes context shapes the manner in which risk is regulated, or what they refer to as 'context shapes content'. 'Regime context' refers to the backdrop of regulation. There are three elements that Hood *et al.* use to explore 'context': the technical nature of the risk; the public's and media's opinions about the risk; and the way power and influence are concentrated in organized groups in the regime. These three pressures are commonly employed explanations in the public policy literature and can be related, to some extent, to a normative theory of regulation as well as to a positive one (Hood *et al.*, 61).

Hood *et al.* derive three separate (but overlapping) hypotheses from these three pressures. The first hypothesis, *the Market Failure Hypothesis*, examines the government's intervention as a necessary one given the technical nature of the risk and the inability of the market to manage the risk effectively without such intervention. The second hypothesis, *the Opinion-Responsive Hypothesis*, examines the extent to which risk regulation is a response to the preferences of civil society. The third hypothesis, *the Interest Group Hypothesis*, examines the role of organized groups in shaping the manner in which a risk is regulated in the industry.

Hood *et al.* use these separate hypotheses to determine the extent to which each of these aspects of context explains the size, structure and style of risk regulation, or what they call 'risk regulation content'. Regulation content refers to the policy settings, the configuration of state and other organizations directly engaged in regulating the risk, and the attitudes, beliefs and operating conventions of the regulators (Hood *et al.*, 21).

Each of the three critical elements of 'regime content' is characterized further through the three elements of a cybernetic control system— information gathering, standard setting and behaviour modification. In this sense control means the ability to keep the state of a system within some preferred subset of all its possible states. If any of the three components is absent, a system is not under control in a cybernetic sense (Hood *et al.*, 23–5). Therefore, in addition to referring to the style, structure and size of the regulatory regime, they refer to the regulatory regime's ability and willingness to gather information, set standards and modify behaviour by way of keeping the regime under control.

The approach to this research is based on the Hood *et al.* framework although it will be applied in a slightly different manner. Hood *et al.* apply the framework to examine the relationship between regulators (government) and the regulated (across society, often non-government). In this book, I will seek to examine how the governments regulated the risks within their own operations as well as those risks outside government. In order to broaden the concept of risk regulation to include this internal dimension I will use the term 'risk management'.[6] I will note one other important deviation from the Hood *et al.* framework. Hood *et al.* use the Wilson (1980) typology to explore the extent to which interests explain risk regulation. I use the Marsh and Rhodes (1992) Networks typology (see also Rhodes, 1997). The reason for the substitution will be explained at the beginning of Chapter 6, the Interests Chapter. Figure 1.1 outlines the approach for this book, listing the three specific hypotheses with the key indicators for each hypothesis captured in parentheses.

In sum, first I intend to describe and analyse the management size, structure and style of both governments' reactions to Y2K according to a cybernetic view of control, essentially populating the right-hand side of Figure 1.1 (Chapter 3). Second, I intend to explore the context that surrounded Y2K through three specific lenses, captured by the left-hand side of the table: the market context (Chapter 4); the public opinion context (Chapter 5), and the organized interests context (Chapter 6). Each of these three context chapters will endeavour to answer the question 'To what extent can the particular sub-hypothesis being tested explain the governments' reactions to Y2K?'

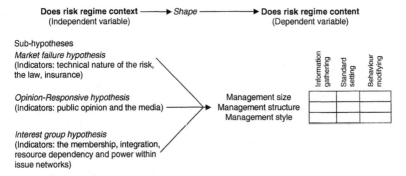

*Figure 1.1* The framework for this book (Hood, Rothstein and Baldwin's (2001) Risk Regulation Regime Framework, adapted)

## Case studies

This book examines comparatively the roles and interplay of the US and UK executives and the relevant legislative committees in relation to Y2K. But in so doing, it also seeks to understand the dynamic between the executives and the government departments and agencies, including the opportunities and constraints that the operational staff faced when attempting to implement the Y2K programme. Therefore, I chose four specific government agencies—two US agencies and two UK agencies—to examine more closely the challenges of implementing the Y2K programme 'on the front line'. The agencies I selected have comparable remits: the Federal Aviation Administration (FAA) and the Civil Aviation Authority (CAA), and the Bureau of Labor Statistics (BLS) and the Office of National Statistics (ONS). All four agencies are heavily dependent upon technology to deliver their respective services. These agencies allow for a comparison across countries and functions (that is, aviation management versus statistics management). In addition, the book examines the roles and interplay of IT service providers external to government, regulated organizations in the aviation sector, the media and public opinion.

## Definitions, clarifications and limitations

For Hood *et al.* regulation means attempts to control risk, mainly by setting and enforcing product or behavioural standards (Hood *et al.*, 1). In this book 'risk management' will be understood in the same manner. Hood *et al.* define risk as a probability, though not necessarily calculable in practice, of adverse consequences (Hood *et al.*, 1). Financial risk is not part of their analysis, nor is business risk, though at times the concepts intersect with their study. In this book, operational risk, and therefore business risk, is critical to the investigation.

Hood *et al.* (2001) define risk regulation as governmental interference with market or social process to control potential adverse consequences to health. For this book the term 'risk management' will be defined as interference with market, social and governance mechanisms and processes to control potential adverse consequences of operational failure.

Therefore, in sum, in this approach all organizations, including individual departments and agencies as well as organizations in the private sector, are considered in the same manner with respect to their ability to manage the potential impacts of Y2K. As the Office of Management and Budget in the United States and the Cabinet Office in the United Kingdom were the offices that issued most of the Y2K directives, one might usefully think of these offices as the central regulators or central managers of the risk. One might claim, however, that the first hypothesis, the Market Failure Hypothesis, is of little import to government departments and agencies because they are not market-driven. While this may in fact be the case, advocates of New Public Management (Hood, 1991) and Reinventing Government (Osborne and Gaebler, 1992) trends would suggest that government departments and agencies ought to have been sensitive to market pressures in the run-up to Y2K. Whether or not they were will form part of the investigation of this book.

Note also that this book examines Y2K from the perspective of the two governments. It focuses largely on government departments and agencies and the government's management of the risk across the infrastructure. While it occasionally refers to private organizations in order to highlight aspects of context, it does not refer to the details of private organizations' management of Y2K. In short, it compares both governments' respective management of Y2K within government and outwith. But this book does *not* compare how a government department managed Y2K with how a private organization managed Y2K, for instance.

For the purpose of a comparative study, some commonly used terms have to be clarified to facilitate comparison and ensure consistency. For this book, the term 'government' refers to government departments and agencies. It does not refer to the respective legislatures nor does it refer to the judiciary. When I refer to the executive or central agencies I refer to oversight agencies in government that include in the United States, for example, the Executive Office of the President (EOP), which included the Office of Management and Budget (OMB) and the President's Council on Year 2000 Conversion. In the United Kingdom it refers to Number 10 Downing Street (Number 10), MISC 4 (the Y2K Cabinet Committee) and the Cabinet Office (CO), which included Central Information Technology Unit (CITU) and its associated agency, Central Computer and Telecommunications Agency (CCTA).[7]

While I occasionally refer to the concept of contingency planning, this research does not include it explicitly. By contingency planning, I mean plans that are implemented in the event of operational failure. These plans are often highly sensitive and therefore difficult for researchers to access and quote from. Nor does the research include the international dimension. The government and the aviation industry, in particular, have a number of international partners that were critical to their Y2K operations but including the international partners would make the research much larger in scope.

Finally, this book is pitched primarily at those interested in issues of public administration and both quantitative and qualitative aspects of risk management, information technology management and infrastructure protection. I make no claim to reinvent the technical environment in the period leading up to 1 January 2000, with an eye to answering the counterfactual *'What if* we had done nothing, would anything have happened?' (a question I receive frequently). As noted, as best as possible, I will let the data speak for itself. Many of my sources have a strong technical bias (for example, IT programmers, technical Y2K reports). What became very clear early on in this research, which began formally in 2001, was that although most people with an IT background agree that Y2K would have caused operational problems had there been no intervention, there was and continues to be significant discrepancy over the degree of impact that it would have had. While I do include alternative views[8] to the prevailing wisdom of the time, this book will not attempt to answer those 'what if' questions.

## What this book argues

In the face of the uncertainty surrounding Y2K the executive level of both governments orchestrated a slow, detailed, expansive and standardized response within departments and agencies. There was little tolerance for anything less than full, demonstrable Y2K compliance, irrespective of cost or level of criticality of system or service. With respect to the national infrastructure both governments organized voluntary fora that included representatives of sectors deemed critical to the functioning of the national infrastructure. The US government's comparatively 'arms-length' approach to industry can be understood through a pluralist lens, whereas the UK government's more interventionist approach with industry can best be viewed through a corporatist one.[9]

Y2K was led largely by an elite group, which included business elite, Congress, the media and some sectors within the IT industry. The initial groundwork for planning was set before it became an issue of general public opinion, which the governments tried to shape rather than follow. Yet contrary to the popular post-1 January 2000 backlash,[10] the technical challenge relating to Y2K posed a serious challenge to the successful management of the national infrastructure. The enormous response to Y2K can partly be attributed to key interdependencies that had not been mapped, noted or managed; poor IT programming and maintenance practices; and growing dependence on decentralized IT.

The enormous response can also be attributed to a clash of competing rationales. An objective understanding of risk, which underpins the traditional rationale for IT risk management, advocates risk identification, segmentation and elimination. More recent interpretations of the risk concept, on the other hand, which were emerging in government risk strategies, see risk as complex, pervasive, multi-faceted and, at times, unmanageable. These two views merged during Y2K: organizations applied IT management tools that commanded an orderly and systematic approach to a risk that was largely understood to have no boundaries. This inevitably led to a massive response. With stability as the ultimate goal, the respective governments attempted to eliminate the risk within departments and agencies and formed practical yet tenuous relationships outwith by way of doing the best they could in a bad situation. In short, quite contrary to Hood *et al.*'s hypothesis that context shapes management, with Y2K, at least, there were times when management—the tools and techniques adopted—shaped context.

## Plan of the book

In addition to this introduction and an appendix, which outlines the research methods, this book contains six chapters. Chapter 2 reviews trends in IT and risk management in both governments and relates these trends to recent debates in the risk literature. Chapters 3–6 deploy the Hood *et al.* framework explicitly: Chapter 3 describes comparatively the Size, Structure and Style of the management of Y2K by the respective executives, statistics agencies and aviation agencies; Chapter 4 tests the governments' reactions against the first of the three hypotheses, the Market Failure Hypothesis; Chapter 5 tests the governments' reactions against the second hypothesis, the Opinion-Responsive Hypothesis; and Chapter 6 tests the explanatory capacity of the final hypothesis, the Interests Hypothesis. Each of these four central chapters starts with a brief description of the hypothesis and with a commentary on the capacity of the framework to elucidate the Y2K story.

The final chapter, Chapter 7, concludes by returning to the central thesis—the extent to which context shaped the governments' management of Y2K. It also outlines the utility of the framework and what this research can contribute to the theory and practice of IT risk management in government and CIP more generally.

I started this research in 2001, only weeks prior to September 11. I sought to understand better what might be described as the 'Y2K phenomenon'. It proved to be a very rich case study. Indeed, while IT commentators, both professional and amateur, frequently offer IT as the universal solution to social and organizational problems, Y2K demonstrates that the 'panacea' can quickly deteriorate to 'organized pandemonium'.

# 2
# Risk—A Contested Concept

This chapter will provide a brief overview of the trends at the turn of the millennium in risk and information technology (IT) management in the two governments. These trends reveal optimism on the part of the governments in question about the capacity of IT to improve efficiency and effectiveness in service delivery, while at the same time reveal only mixed success in IT projects. Indeed, the—at best—moderate success of many technology projects helped to generate increased interest in managing risks more effectively. In some ways, therefore, the fates of IT management and risk management were joined.

The dialogue about risk management in government, however, is symptomatic of a larger debate in the social sciences about the concept of risk. The second part of the chapter, and by far the part that will consume most of our time, will review this debate more closely. It will start by examining the traditional concept of risk, which largely underpins many risk practices in IT management. A positivist and reductionist understanding of the world underpins the approach. From this traditional view, risk is largely understood to be a negative concept; people seek to identify, segment and eliminate it. The chapter will then consider challenges to this traditional view that have emerged from the fields of psychology, sociology and anthropology, respectively. This multi-disciplinary analysis of the concept will demonstrate that there are many ways to interpret and understand the concept of risk. Each approach makes different assumptions about the nature of risk, which in turn has an impact on the tools and mechanisms required to manage it. While taking multiple views into account almost certainly enriches our understanding of the concept, it also introduces potentially incompatible notions of risk that have to be managed and trade-offs that have to be decided. While taking one view is potentially narrow, taking all views is potentially unwieldy.

## Trends in IT and Risk Management in Government

IT projects have a precarious legacy (NAO, 2004b). Margetts noted a surge of enthusiasm at the Cabinet Office in the mid-1990s under Deputy Prime Minister Heseltine that had reversed a trend of cutbacks that had been occurring at the governments' IT agency, CCTA, throughout the 1980s and early 1990s (Margetts, 1999, 45). The Conservative Government issued a green paper, *government.direct*, that foresaw investment in IT delivering a more integrated government service, which included reduced costs and more efficient service delivery (Cm 3438, 1996, 1). Much of the initial increased expenditure in IT, however, did little to offset the view that IT projects frequently start as ambitious 'reinvention plans' but end-up over budget, late and with a greatly diminished capacity compared to what they were supposed to do in the first place. There are numerous examples of IT projects that are viewed as failures on these grounds. (See Heeks, 1999, 49–55 for examples).

Following the Conservative government's initiative, the Labour government placed IT in the heart of the Modernization Agenda. *Modernising Government* includes a chapter on the 'Information Age' (1999, pp. 44–53), in which it describes improved public service through such practices as electronically joined-up government service, e-commerce and access to government documents and services through IT. This IT advocacy was quickly followed up by *Successful IT: Modernising Government in Action* (Cabinet Office, 2000), which documents numerous IT failures, shortcomings and recommendations. The report concludes, 'improvement in the public sector will require effective use of information technology' (5). Indeed, even the government's Y2K post-mortem is entitled *Modernising Government in Action: Realising the Benefits of Y2K* (2000). Perhaps most noteworthy was Prime Minister Blair setting a 2005 target by which all departments and agencies would have their services available online.[1] In addition, the government set up central structures, such as the e-envoy and Office of Government Commerce (OGC) as a way of facilitating information-sharing among departments and agencies as well as setting standards for government. OGC has encouraged departments to integrate IT projects into broader departmental goals and initiatives and helped departments to achieve 'value for money' in their IT projects (OGC, 2004), through the implementation of risk management plans in collaborative projects; these plans include more systematic monitoring, reviewing, managing and communication of risks (HM Treasury and OGC, 2005, 5–10). NAO reports have been generally supportive of the role the OGC has played (NAO, 2004b, 1).

The NAO has been less supportive of departmental efforts to realize the potential benefits of IT (NAO, 2004b, 1). The NAO Value for Money Audits from a few years back that focused on mismanagement, such as *The Cancellation of the Benefits Payment Card* (NAO, 2000b) and *The UK Passport Agency: The Passport Delays of Summer 1999* (NAO, 1999c) have given way to audits that criticize government for failing to tap into IT as a potential resource to improve services. *Better Public Services through E-government* (NAO, 2002a) and *Government on the Web Parts I and II* (NAO, 1999d, 4; 2002b, 5) all criticize departments for failing to build-up IT capacity, with respect to the technology itself as well as the human resources required to use and manage it. In *Progress in Making E-Services Accessible to All: Encouraging Use by Older People* (NAO, 2003, 10) the NAO also criticizes government for failing to reach-out and advertize the potential to groups that traditionally have not used online services as much as others.

In some ways IT management and risk management have shared a common fate in the UK government, and indeed the subjects often overlap, particularly because, as noted, IT projects frequently had such a bad track record of not meeting their goals (Cabinet Office, 2002, 2). Like IT management, risk management has also increased its profile at central agencies as well as in government departments and agencies. The Strategy Unit at the Cabinet Office (Cabinet Office, 2002), OGC (2001) and the Treasury (HM Treasury, 2004; 2005; with OGC, 2005), for instance, have expanded their interest and often their oversight role in working with departments and agencies to identify and manage risks.

Again, like IT, central agencies are encouraging greater integration of risk management practices into 'business as usual'.[2] These centralized efforts were partly by way of bringing some commonality to a field that was quite disparate. The UK Interdepartmental Liaison Group on Risk Assessment (UK-ILGRA) noted a variety of practices and assumptions across government, though noted the overall tendency for precautionary, 'better to be safe than sorry' thinking (HSE, 1998). ILGRA also noted a tendency towards quantitative analysis, which marginalized important risk detection methods. The NAO's reports on risk have also noted across government a variety of practices, policies and degrees of integration, ranging from active and formal risk management policies to non-existent ones (NAO, 2000a, 12–17; 2004a, 14–18).

The Strategy Unit has tried to 'open up' departments' perspectives on risk detection and management, encouraging the use of non-quantitative tools and broader risk management strategies, including, for example, risk communications plans, early stakeholder consultation and drawing a distinction between strategic programme and operational risks. The Strategy Unit also highlights the challenge of developing trust among the public at a time when trust is decreasing among the public towards established institutions (2002, 3). For the Strategy Unit (2002, 10) and the Treasury (2004, 13) risk management is an ongoing process that is embedded in projects and programmes, involves ongoing communication and learning and occurs in a particular (potentially crowded) context (for example, economic concerns; stakeholder concerns). Responsibility lies with departments and agencies and central agencies play a supportive role of coordination and feedback (2002, 23). Notably, also, risk can also be seen as an 'opportunity' to be encouraged.

The Strategy Unit report, however, fails to acknowledge important contradictions in its report and the related challenges with implementing it. It emphasizes departmental level responsibility though it continues to build-up central oversight bodies at the Cabinet Office and the Treasury that will challenge if not undermine 'ownership' at the departmental level (2002, 23). It encourages opening up dialogue on risk in order to gain multiple perspectives but is silent on what to do in the event of incompatible risk views, such as those of which Cultural Theorists would warn (Douglas, 1982; Hood, 1998). It notes it wants to create a 'risk-taking' culture (2002, 4) though provides little detail on how to encourage such a reversal against type. Moreover, it notes one of its principal aims is to eliminate 'surprises' (2002, 1), which hardly seems consistent with a risk-taking culture. Nor does it address the issue of balancing the need to refer and consult with the need to keep costs down. Yet both are noted as important objectives.

PRINCE2 (Projects in Controlled Environments),[3] a tool developed collaboratively between public and private sectors, is a commonly used project management method in practice in the UK public and private sectors. PRINCE2 advocates a form of risk management that should occur throughout the entire systems development life cycle (SDLC).[4] In PRINCE2 risks must be managed in a structured way. Risk analysis (that is, identify, evaluate, action) is followed by risk management (that is, planning, resourcing, managing). PRINCE2 advocates risk logs (or risk registers and risk profiles), which (1) identify risks; (2) assess the risks; (3) name those responsible for managing the risk; and (4) indicate the status of the risk. It also acknowledges different degrees of acceptance of risk. Risk management efforts, however, must be analysed for their monetary worth; it recommends cost-benefits analyses (2002, 87).

PRINCE2 has specific assumptions. In PRINCE2, risks are viewed almost exclusively as negative. It assumes risk can (and will) be articulated and managed. It has a top-down bias. It does not acknowledge competing risks or incompatible solutions though it does acknowledge complexity in the number of players involved in projects. In the Y2K context, for example, PRINCE2 would run into trouble. It emphasizes controlling the environment, yet with significant interdependencies within and outwith government, this control is extremely difficult to accomplish. Such an endeavour also runs head long into one of its other tenets—trying to manage costs. Trying to control a practically uncontrollable environment by implementing redundancies and contingency plans, for instance, can quickly become expensive. Also, it seems to have an anti-government bias. Its use of bureaucratic, for instance, is pejorative. Ironically, PRINCE2's approach to risk management would almost certainly feel at home in Weber's Bureaucracy: methodical; detailed; specialized; and controlled.

We see equally conflicting messages about risk and IT from the US government. The US Government's recent push for increased use of IT date to the Reinventing Government initiatives that were led by Vice President Al Gore shortly after President Clinton's election (Gore, 1993). IT was envisioned as a means of improving service and cutting costs. The CIO Council was established in 1996 by Executive Order 13011. The Council serves as the principal interagency forum for improving practices in the design, modernization, use, sharing and performance of federal government agency resources. Every major department has a member on the Council. It has three subcommittees—Best Practices; Enterprises Architecture Committee; and the IT Workforce/HR.[5]

In 2002 the US Government enacted the *E-Government Act*. The Act provided for OMB to strengthen its oversight and coordination role across government by, for example, establishing the Office of E-Government. The purpose of the Act is to promote better use of the Internet and other information technologies to improve government services and enhance government opportunities for citizen participation in government. In February 2002 the US government launched its *E-Government Strategy* (OMB, 2002). The strategy advocated better 'business architecture' across government, which meant in part a greater integration and (when necessary) standardization of technology across government. Such integration, it argues, will lead to the elimination of redundant and overlapping agency programmes. The strategy noted, however, that its initial research unearthed several barriers to innovation, chief among them: organizational culture; existing architectures; lack of trust; lack of resources; and stakeholder resistance (OMB, 2002b, 2).

On the balance, the GAO has been supportive of the progress that OMB has made on this front, although it did note in a December 2004 report that OMB had failed to initiate activities in crisis management, contractor innovation and federally funded research and development (GAO, 2004c). GAO has been more consistently critical, however, of departmental advances in enterprise architecture, which it noted hardly progressed at all between 2001 and 2003, according to GAO's *Maturity Scale* (the GAO's multi-step method for charting progress in this area) (GAO, 2004b, 2–4).

Neither the EOP nor OMB has laid out explicit risk management strategies in the same way as the CO has. That noted, John Graham, Administrator for the Office of Information and Regulatory Affairs at OMB, articulated a White House view on risk management in a speech to the International Society of Regulatory Toxicology and Pharmacology (Graham, 2002). Like the CO view, the understanding of 'risk management' is much broader than traditional science-based views and in fact casts aspersions towards an overly risk-averse culture. Graham made five related conclusions. First, precaution is necessary and useful but it is also subjective and subject to abuse by policy-makers for trade purposes or other reasons. Second, scientific and procedural safeguards need to be built into risk management decisions that are based in part on precaution. Third, adoption of precautionary measures should be proceeded by a scientific evaluation of the hazard and, where feasible, a formal analysis of the benefits, risks and alternative precautionary measures. Fourth, concerns of fairness, equity and public participation need to be reflected in risk management. Finally, the set of possible precautionary measures is large, ranging from bans to restrictions to education. The mix of the appropriate measures is debatable.

Graham (2002) advocates more of a risk-taking culture. He warns against the precautionary principle, for instance, arguing that too cautious an approach in the face of uncertainty can undermine innovation and job creation. Indeed, he advocates more transparency in the assumptions that are embedded into scientific research, which he suggests are often too conservative. He notes that 'sometimes risks prove far worse than expected; other times predictions of doom simply do not materialize'.

Again, however, we see conflict between the centrally held view of risk and the view of risk understood by centres of technology policy and practice. The National Institute of Standards and Technology (NIST, part of the US Department of Commerce) promotes the US economy and public welfare by providing technical leadership for measurement and standards in infrastructure. In 2001 NIST issued the *Risk Management Guide for Information Technology Systems*. Like PRINCE2, NIST advocates integrating risk management into the SDLC. 'Risk' is understood as entirely negative (NIST, 2001, E-2) and the report as a whole deals almost exclusively with security issues.

In the NIST report, risk management strategies are described as flow-charts (2001, 9); they are predominantly methodical and step-by-step (2001, 31). Security is divided into three tiers: management; operational; and technical. It advocates the use of short-hand risk tools in which the likelihood and impact of the risk are captured on a three-by-three matrix in which risks are defined as high, medium or low risk (2001, 25). According to the report, risk management strategies should be viewed and advocated based on cost-benefit analyses, in which costs and benefits are articulated in dollars (2001, 37–8). Because of the emphasis on security, people are viewed as 'threat sources' (2001, 15).

While the approach is detailed yet vast in scope it shares many of the biases of PRINCE2, in which risks can be captured and managed. While it is tempting to view the report as a reaction to 9/11 (the report was issued in October 2001) in fact IT security has long been of considerable concern in the United States. I note a few pre-9/11 examples of infrastructure protection, risk and continuity planning here: *The Computer Security Act* (1987); Federal Executive Branch Continuity of Operations (Federal Preparedness Circular 65, July 1999); Enduring Constitutional Government and Continuity of Government Operations (Presidential Decision Directive, October 1998); Critical Infrastructure Protection (Presidential Decision Directive 63, May 1998); as well as federal response plans from the Federal Emergency Management Agency.

In 2002 the US government also enacted the *Federal Information Security Management Act* (FISMA), which was part of *The E-Government Act* (2002). Its goals include development of a comprehensive framework to protect the government's information, operations and assets. FISMA requires agency programme officials, Chief Information Officers and Inspectors General to conduct annual reviews of the agency's information security programme and report the results to OMB. The OMB uses the data to assist in its oversight responsibilities and to prepare its annual report to Congress. Its 2004 report concluded that anywhere from 76 to 85 per cent of government systems were passing tests on three of its four measures—Security and Privacy Controls; Built-In Security Costs; and Tested Security Controls. Contingency plans, however, continued to lag behind at 57 per cent (OMB, 2005, iii).

In sum, while the IT practice largely treats risk as a negative concept, which can largely be defined, captured and managed, if not eliminated, emerging views of risk are broader in their interpretation. The central agencies of OMB and CO, for instance, have encouraged greater stakeholder consultation to ensure multiple interpretations of risks, an aversion to precautionary approaches, and even risk as presenting opportunity rather than merely a potential loss. These conflicting views of risk in fact reflect a debate that has been occurring in the social sciences about the concept of risk, which can largely be understood as encompassing four distinct rationalities, discussed in the next section.

## Risk Rationality

Up until the 1980s the study of 'risk' was dominated in both the United States and United Kingdom by scientists, engineers, economists and decision analysts. Their views are overwhelmingly influenced by a rational actor paradigm (RAP) (Jaeger *et al.*, 2001, 19–22), in which risk is largely understood as an objective condition with a rational/individual bias. This section seeks to summarize this dominant view of risk, including its applications and its related strengths and weaknesses. The review moves then to consider the important contributions and challenges to the risk debate posed by Psychology (Psychometrics), Sociology (Critical Theory and Systems Theory) and Anthropology (Cultural Theory). As an organizing framework I will use Renn's Risk Rationality Diagram (below) to consider the traditional and alternative approaches. This review of the risk literature draws significantly from the risk reviews of Adams, 1995; Jaeger *et al.*, 2001; Taylor-Gooby, 2004; and Zinn, 2004. Jaeger *et al.* in particular have had a considerable influence on my thinking in this subject.

Figure 2.1 below organizes a discussion about risk and risk research by competing rationalities. The first and dominant view is the rational actor paradigm (RAP) (Box A). For this view, risk is objective and understood through the lens of an individual. Box B, largely the domain of psychologists, views risk through an individual's lens but assumes that risk is a subjective (that is, personal, intimate) construction. Box C, largely the domain of the sociologists, some natural scientists and many business schools, also assumes that risk is objective but views such risk through structural or organizational settings. Box D, largely the domain of the sociologists and anthropologists, understands risk to be constructed but through a structural setting. The following discussion examines each of the four.

### Objective/Individual

*A Brief Overview of the Rational Actor*

The rational actor is one who when confronted with a decision deliberates on the feasibility of alternatives, desirability of outcomes and the causal

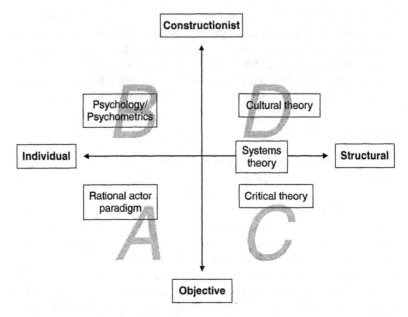

*Figure 2.1*   Risk Rationality Diagram
*Source:* Based on diagram presented by Renn, at the Social Contexts and Responses to Risk Inaugural Conference, January 2005, at the University of Kent at Canterbury. Updated version can be found in Renn, 2008. Reproduced with the author's permission.

relationship between the two. More formally, Simon (1954) described rational choice as follows (as cited in Pidd, 2003).

The most advanced theories, both verbal and mathematical, of rational behaviour are those that employ as their central concepts the notion of:

1. a set of alternative courses of action presented to the individual's choice;
2. knowledge and information that permit the individual to predict the consequences of choosing any alternative; and
3. a criterion for determining which set of consequences he prefers.

In these theories rationality consists of selecting that course of action that leads to the set of consequences most preferred.

A simple illustration of such a decision problem would be in deciding between taking the train or the bus to work where the consequence of taking the train is paying £3 compared with £1 for taking the bus. Using cost as the sole criterion for the decision we chose the bus as the preferred alternative as the consequence will be a savings of £2.

In practice, things are seldom so simple. We often have many alternatives from which to choose, and decisions are not always 'one-offs'. They can be sequential too. As a result, we can be faced with having continually to make decisions, with the choices we make today affecting the alternatives and consequences of the future. In addition, the valuation of consequences is not always straightforward. We can often be confronted with multiple criteria against which to evaluate alternatives. Moreover, the relationship between decisions and consequences is not always known so precisely. One of a number of possible outcomes can result from a decision.

If we wish to extend this analysis to a group-making exercise, such as an organization or society, we assume that we can combine the utilities of individuals such that we have one utility function against which we can compare alternatives. Likewise we assume we can assign probabilities to events, uncontroversially. Organizations function as a unitary actor. There is one right answer. Given the resources, everyone—thinking rationally—will arrive at the same conclusion.

### Implications of the Rational Actor Paradigm to Risk

Starr (1969) and Lowrance (1976) are among the most significant early contributors in this particular approach. *Probability multiplied by consequence* is perhaps the most famous definition of risk. Within this view, technical risk analyses are assumed to be able to reveal, avoid and/or modify the causal agents associated with unwanted effects.

Until the early 1980s this understanding of risk was largely uncontested. In 1983, the Royal Society in the United Kingdom described risk as a probability that a detrimental event would occur during a stated period of time or result from a particular challenge (cited in Adams, 1995, 8). In short, risk could be calculated by combining probabilities. In this context, 'detriment' was defined as a numerical measure of the expected harm or loss associated with an adverse event. 'Detriment' was usually the integrated product of risk and harm and was often expressed in terms such as cost in money, loss in expected years of life or loss of productivity. The National Research Council in the United States came to conclusions similar to those in the Royal Society (Adams, 1995, 8).

There are specific tools that lend themselves to a RAP approach to risk. Probabilistic risk assessments (PRAs) offer a method for analysing and predicting the failures of complex technological systems. Users assess systems failures by reducing the systems to their operating component parts. Estimates of a systems failure are often based on 'fault tree' and 'event tree' methods, for instance (Jaeger *et al.*, 2001, 90).

The example below (Figure 2.2) is an event tree analysis taken from the United Kingdom's Railway Safety and Standards Board (2005). It is a high-level example; it does not include the numerous calculations that lie beneath each event. It is meant only to be illustrative. It estimates that 'an event' (passenger train derailment) will occur 9.8 times per year. Given the derailment, it then makes further estimates about related events (for example, the likelihood that there is a collision with a train on an adjacent line). If one follows the event tree to its conclusion, the event tree exercise estimates 2.76 people will die each year as a result of passenger train derailments.

This approach to understanding risk has been criticized on a number of grounds. First, there are practical problems with obtaining data for these models. To start, the interaction between human activities and consequences is more complex and perhaps subtle than the average probabilities captured by most risk analyses. Moreover, when data is unavailable for these models—and it often is unavailable when we are exploring rare events (for example, acts of terrorism)—the data is often estimated. Estimations embedded at several levels of complex models will undermine the overall validity of the model. Finally, data is often collected and models are built on past experiences, as is usually the practice in actuarial science. These models will fail to predict new or rare events because the assumptions of the past do not necessarily hold. Second, the institutional structure of managing and controlling risks is prone to organizational failure, which may increase actual risks (Jaeger *et al.*, 2001, 86).

From a normative standpoint this approach embeds key assumptions. To start, complex technological systems are accessible to detailed human comprehension and that a reductionist approach is the best way to understand the systems. Motivation, organization and culture are ignored (Jaeger *et al.*, 2001, 91). Moreover, the approach upholds the privileged position of the one who designed the model—the expert. Finally, risk minimization is not necessarily the only end in mind; equity, fairness, flexibility and resilience are also plausible and potentially desirable goals (Jaeger *et al.*, 2001, 86).

## Constructionist/Individual

### Psychology and Psychometrics

The psychometric paradigm draws on the work of cognitive psychologists such as Slovic (1992) to conceptualize risks as personal expressions of individual fears or expectations. In short, individuals respond to their perceptions whether or not these perceptions reflect reality.

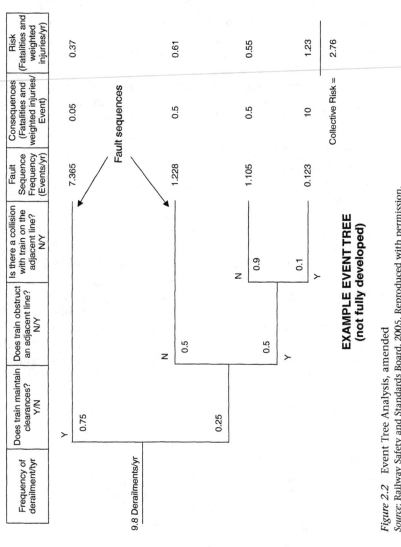

*Figure 2.2* Event Tree Analysis, amended

*Source:* Railway Safety and Standards Board, 2005. Reproduced with permission.

The psychometric approach seeks to explain why individuals do not base their risk judgements on expected values, as RAP advocates would suggest (Jaeger *et al.*, 2001, 102–4). The approach has identified several biases in people's ability to draw inferences. Risk perception can be influenced by properties such as perception of dread, personal control, familiarity, equitable sharing of both benefits and risks and the potential to blame an institution or person. It can also be associated with how a person feels about something, such as a particular technology. People also believe that events that they have experienced personally are more likely to occur than aggregate data suggests. Information that challenges perceived probabilities that are already part of a belief system will also be either ignored or downplayed. Finally, there are different classes of meaning. Risk can be understood as a random threat that triggers a disaster; an invisible threat to one's health; a balancing of gains and losses; or something to be actively explored and desired.

This approach to risk still has limitations. It suffers from the same micro-orientation as RAP. It assumes that an individual will act on his or her subjective estimates of consequences and probabilities. It is not clear if an individual will pursue a strategy to verify or validate his or her perceptions before acting on them (Jaeger *et al.*, 2001, 107). Moreover, lay judgements of risk are seen as multi-dimensional, but a strong distinction between the subjective 'popular level' and the objective 'expert level' is maintained (Taylor-Gooby, 2004, 7). Methodologically, psychometrics has also been criticized for gathering data through questionnaires, in which the issues are already predefined (Taylor-Gooby, 2004, 6–7).

*Cost Benefit Analyses*

Cost Benefit Analysis (CBA)[6] is likely the most common decision analytic tool. CBAs need not necessarily originate from the RAP paradigm; they could be understood as personally constructed. They aim to create a consistent and logically sound model of a person's or institution's knowledge and preference structure (Jaeger *et al.*, 2001, 79). The tool assumes 'positives' and 'negatives' can be articulated, compared and judged in a single measurement: usually dollars. They require participants to engage in multiple acts of conversion, assigning monetary values to such entities as human lives, human morbidity and a range of harms to the environment. American government for instance assigns such values on the basis of private 'Willingness-to-Pay' (WTP). For example the Environmental Protection Agency values a human life at about $6.1 M;

the figure is the product of many studies of actual risks in the workplace, the housing market and the market for consumer goods, attempting to determine how much workers and others are paid to assume mortality risks (Sunstein, 2007, 200–1).

Claims that decision analytic procedures such as CBAs always produce outcomes that decision-makers prefer are suspect. People have blind faith in numerical analysis and computer models; in fact, such analyses depend on the ability and willingness of the decision-makers to transfer their preferences into numbers (and often dollars) (Jaeger *et al.*, 2001, 90). These processes are also subject to the bias and potential manipulation of the analysts when they present data (Jaeger *et al.*, 2001, 81–2). The model fails also to include the more subtle dynamics in decision-making, such as strategic reasoning, power plays, interests and institutional responses, for instance (Jaeger *et al.*, 2001, 82). Dietz and Stern (1995) note also that the relatively complex mathematics does not correspond with what we know about human behaviour. Rather, people are good at pattern recognition, classification and applying rules of thumb.

Finally, those in psychometrics have challenged CBAs on some of their suspect outcomes. For instance, when the questions are framed as potential losses rather than gains, even if the cases presented are exact mirror images of the same choice situation, they produce conflicting results. In a group, community or institutional setting, varying degrees of risk aversion make it difficult to determine a collective value for risk. Also, chains of probabilistic events can increase the overall uncertainty of the final outputs to such a degree that articulating with any reliability the costs and benefits is extremely difficult (Jaeger *et al.*, 2001, 116).

*Risk Communication and the Media*

Most risk communication/media research has originated from the field of psychology and social-psychology and we will therefore consider the relevant literature here. Much like the risk research in general, media research on the question of risk focused originally on objectivity, rationality and accuracy of media coverage (Freudenberg *et al.*, 1996; Wilson, 2000). Researchers noted that many people base their perceptions about risk primarily on information presented in the media (Fischhoff, 1985; 1995; Kitzinger and Reilly, 1997). Yet researchers also noted the media's propensity to report the dramatic over the common but more dangerous (Soumerai *et al.*, 1992), its tendency to senzationalize (Johnson and Cavello, 1987) and its dependence on experts without having expertise itself to counter-act the claims it receives from the experts (Freudenberg *et al.*, 1996). This approach loaned itself to the view that distinguishes between expertise and ignorance; it is concerned with improving communication by way of ensuring that lay mental models, for instance, correspond more closely with those of experts (Fischoff, 1997, as cited in Taylor-Gooby, 2004, 5).

However, more recent developments in risk research challenged this approach. On the one hand, the fundamental assumption that the media should support the public in making adequate judgements by giving objective information was challenged by the problem that often objective knowledge is not available (Adams, 1995; Kitzinger, 1999; Murdock *et al.*, 2003). On the other hand, the widely disseminated assumption that media reports have a determining influence on public risk perception was challenged by the observation that the 'subject' has a relatively more active role concerning the interpretation of and response to risk.

This view is not altogether 'new'. One can see evidence of it in Downs (1972, 39–40) when he argues, for instance, that issues go through peaks and troughs of interest as people become aware of and alarmed by the 'evils' of certain issues but then either become bored with the problem in question or realize how difficult and costly the solutions are.

More recent contributions along these lines have come, for instance, from Wahlberg and Sjoberg (2000) who note that the media's influence is too often taken for granted when in fact much of the evidence points the other way—that media are probably not a strong causal factor of (especially not personal) risk perception. Risk perception may be affected by the media but the effects are lessened by impersonal impact. Moreover, general risk perception is more easily changed than personal risk perception. Finally, it is not conclusive that risk perception changes behaviour.

Similarly, Mutz and Soss (1997) note that the media raises people's perception of the salience of a subject in the community but is much less successful in changing people's mind on a particular subject. Similarly, Atwood and Major (2000) note that people do not think of themselves as being as vulnerable to risks as others are. Indeed, some suffer from cognitive dissonance; they are unrealistically optimistic, ignoring the news and denying personal vulnerability. In other areas of research, it has been suggested that most individuals gain information from a variety of sources, not just the media (Verba and Nie, 1972), including other individuals, government organizations and advocacy groups.

Studies that compare media coverage at different points in time tend to show that the social and political context and their changes over time are essential for understanding risk reporting (Kitzinger, 1999, 59). Zinn concludes from the literature that research on the framing of risk-perception by the media can only be fully understood by simultaneous analysis of the context in which such risk-reports are embedded and a carefully constructed ethnographic analysis of the individuals' 'embeddedness' in cultural and social contexts and biographical experiences (Zinn, 2004, 16–17).

### Structural

The remaining risk rationalities that we will consider posit the social context at the heart of the interpretive exercise. Here, institutional arrangements not only matter, they hold the key to understanding risk and our responses to it. Advocates of Critical Theory—a neo-Marxist approach to social arrangements—would warn that RAP-based approaches would concentrate power into the hands of a social and political elite, and that this elite would manipulate populations with this power. Systems Theory, Risk Society and Normal Accidents Theory interpret modern societies as increasingly complex to the point of undermining cause and effect relationships that RAP analysts would hold at the centre of their analysis. For these theories, social complexity means accidents are inevitable. To suggest otherwise is simply hubris. Finally, Cultural Theory's approach to understanding risk is entirely constructed through the institutional arrangements in which one finds oneself. In this section on Structure, meanings are problematized, contested and at times, even ephemeral.

### Objective/Structural

#### Critical Theory

For critical theorists, the rational actor paradigm embodies a positivistic approach to the natural sciences that presumes that the natural sciences hold the exclusive model for all knowledge. This presumption leads to inadequate reflection upon the appropriate scope and application of such science. From this point of view, the tools and methods derived from RAP leads to administrative structures that are insensitive to true human needs; the approach reduces humans to mere objects to be manipulated. If society is not sufficiently vigilant, power can be concentrated into the hands of a small group of specialists in social and political elite apparatuses (Jaeger *et al.*, 2001).

The Railway Safety and Standards Board Event Tree Diagram cited in the Rational Actor section is a typical example. In Normal Accidents Theory (noted below) advocates would assert that such a complex system cannot be so readily understood, charted and controlled. In Critical Theory, however, the objection is somewhat different. First, critical theorists would reject this calculation in principle. They would argue that the calculation has been developed by specialist administrators, working in an office, perhaps far from the relevant communities. Moreover, the calculation is an abstraction. It seems logical, rational *and* clinical. But what is it actually implying? It implies that derailments—and therefore the related deaths—are inevitable (and perhaps even acceptable, to a degree) when using this technology. Moreover, in presenting the information in this abstract and reductionist way—calculations, ratios, data, inputs, outputs—it marginalizes the value of human life. Second and relatedly, critical theorists would want access to information that is not presented. Critical theorists would wonder to what extent this issue reinforces socio-economic inequalities. Who benefits from these operations, for instance?

Habermas, influenced significantly by social and moral atrocities of the twentieth century and the chief architect of Critical Theory, encourages individuals to reflect upon whether and how their actions may be inconsistent with their interests. He argues that lay views must be valued, not strictly the views of experts or social elites (Jaeger *et al.*, 2001, 238). His Communicative Action (1984) theory is inherently purposive; its aim is to reach uncoercive consensus among numerous competing views (Jaeger *et al.*, 2001, 238). Critical Theory takes seriously that individuals living together must intentionally discuss preferences, interests, norms and values in a rational way.

### The Precautionary Principle

In some respects, the Precautionary Principle might be thought of along the same lines as Critical Theory. One definition of the precautionary principle states that the lack of scientific certainty is not a sufficient reason to delay a policy if the delay might result in serious or irreversible harm (Jaeger *et al.*, 2001, 115). Because the science is often contested (for example, greenhouse gases or climate change), the processes enacted to support these precautionary approaches are often considered more inclusive of alternative views of issues—they do not simply rely on hard data because the hard data is not always reliable or available.

Despite its popularity, many have challenged the precautionary principle, and in particular what might be described as inconsistencies and contradictions at the heart of the principle. First, while the precautionary principle is widely used it has many definitions. Kheifets *et al.* (2001) have noted three quite different trends in its definition and subsequent application, ranging from modest attempts to understand the dangers associated with certain practices to outright bans of practices that potentially create problems. There is also considerable variation in where the burden of proof should lie in determining which practices should be stopped (for example, should environmentalists have to prove that a practice is potentially dangerous? Or should industry have to prove it is not?).

Second, assuming one uses one of the stricter definitions of the precau tionary principle in which risk taking is stopped or at least significantly curtailed, it is often cited as a costly approach to risk management (Burgess, 2004; Jaeger et al., 2001, 115; Sunstein, 2005; 2007). To these thinkers, estimating the probability and consequences of risks may involve subjective judgement, but this is far less costly than assuming that all risks have the same probability and magnitude. And while the precautionary principle is often associated with understanding issues at greater depth and from multiple views—a 'let's proceed with caution' attitude rather than a risk-taking one—Sunstein argues that the advo- cates of the precautionary principle too often view risks in isolation, and neglect the risks *caused* by precautionary approaches. The decision to assume a precautionary approach in the development of GM foods, for instance (as we have seen in many parts of Europe), may foreground concern over the long-term impact that could result from genetic modi- fication of parts of the food chain, however, it neglects the potential gains of such intervention, especially for relatively poorer countries for whom the benefits of bumper crops could be life-saving.

Finally, despite best efforts to create a more inclusive approach, whom to include, what is included on the agenda and what is considered a viable alternative are still a hotly contested political debate. They are likely to be reflections of existing power structures (and struggles) in society.

## Constructionist and Objective/Structural

### Systems Theory

Systems Theory draws from Luhmann (1993). Luhmann argues that systems are held together by social norms (Luhmann, 1993, 3) but concedes that these norms are contingent and transitory. The environment is constantly in flux: actors enter and exit the systems, and the systems are constantly under threat of dissolution.

According to Luhmann, risk is not the opposite of safety but rather the opposite of danger, which is a threat from the environment (Luhmann, 1993, 21–2). Risk indicates that complexity is a normal aspect of life (Luhmann, 1993, 23); it is an effort to discuss and control an unknowable future that allows actors to internalize the possibility of experiencing an outcome other than the one hoped for.

The fragmentation of formerly common cultural elements (such as language, value, basic knowledge) disables each individual member of each system from meaningful communication with members of another system beyond the exchange of services and products (Jaeger *et al.*, 2001, 207). Luhmann notes the increased interest in risk at the same time as increased specialization in systems (Luhmann, 1993, 28). In short, there is a paradox. The more systems evolve and specialize, the more critical it is for communication and coordination between these systems. Yet, at the same time these systems become more self-referential and unable to communicate between themselves.

In this setting, there is no risk-free behaviour. Interdependence and complexity means that there is no guaranteed risk-free decision (Luhmann, 1993, 28). Risk strategies are necessarily complex and in themselves risky. These systems pose a challenge to the RAP: there are too many systems and each system is too complex and interdependent for rational actors to know the eventual outcomes of their decisions and actions. Moreover, risk-taking systems can produce risks for adjoining or interdependent systems. As a result, risk produces pathways for conflicts between systems.

The theory provides little guidance about how to manage risk. Japp (1996; 2000, as cited in Zinn, 2004, 17) argues that neither partial rationality of selected groups nor the public interest in general should predominate. Rather, a combination of both is required. In short, suboptimal solutions become acceptable for the advantage of public welfare. Trust is a central issue in Japp's considerations. In this view, trust is needed to generate the readiness for risk-taking. The problem is that in several areas there is no possibility to learn by trial and error. Where catastrophes are possible long term learning by trial and error is not an option. Under such conditions the ability to act can only be protected by trust (Zinn, 2004, 18).

*Risk Society*

Social systems theory is closely related to the role of risk and uncertainty in modern societies, as articulated in Beck's (1992) *Risk Society* (Jaeger *et al.*, 2001). Beck defines risk as a systematic way of dealing with hazards and insecurities induced and introduced by modernization itself (Beck, 1992, 21) and argues that modern science and technology have created a risk society in which the production of wealth has been overtaken by the production of risk (Beck, 1992, 19–20). The creation and distribution of wealth has been replaced by the quest for safety. Progress has turned into self-destruction as an unintended consequence through the inexorable and incremental processes of modernization itself. Beck distinguishes modern risks from older dangers by their scale and invisibility, and the need for experts to detect them, despite their limited ability to do so.

Risks, as opposed to older dangers, are consequences, which relate to the threatening force of modernization and globalization. Risks brought on by modernization go beyond national borders (Beck, 1992, 41–2). They also challenge cause and effect relationships; every act potentially causes risk and therefore no act causes risk (Beck, 1992, 33). At least, it is almost impossible to assign responsibility for generating risks.

Even as risks become more global, transcending national borders, risks simultaneously become more intricate and personalized. People have their own views of risk based on their own experiences and rationales. Social, cultural and political meaning is embedded in risk; there are different meanings for different people (Beck, 1992, 24). The natural scientists' monopoly on rationality is broken, and gaps emerge between scientific and social rationality (Beck, 1992, 30).

The risk society also problematizes the notion of time and generations. Society makes decisions today that generate risks and consequences for future generations. The decision-makers from previous generations cannot be held to account, and future generations cannot be consulted.

Finally, some groups are more affected than others by the growth of risks—the poor are unable to avoid certain risks because of their lack of resources (Beck, 1992, 23). Yet some risks affect the rich and poor in similar ways (for example, smog, radiation). In this sense, the risk society is different from previous views of society that were defined by class structures. Indeed, in the risk society, the best educated and the most well-off are most aware of risks but not aware enough to protect themselves sufficiently such that they become anxious without being able to reconcile or act upon their diversity. All of these problems require political, not technical, solutions.

The concept of risk is linked to reflexivity because anxieties about risks serve to pose questions about current practices (Beck, 1992, 21). Society—as the creator of risk—becomes an issue and a problem in and of itself that only society can deal with. The awareness of the global nature of risk triggers new impulses towards the development of cooperative international institutions, for instance. As a result, the boundaries of the political come to be removed, leading to worldwide alliances.

By way of partially offsetting the negative consequences of the risk society, Beck argues for the institutionalization of conflict as a means of challenging dominant views. He advocates the right to criticize one's employer, for instance, and maintaining a strong and independent court system and media.

Researchers have challenged the theory on numerous grounds but particularly its macro-approach. As noted, Hood *et al.* contend that the macro-view fails to identify the variety in risk regulation that exists across countries and even within organizations (Hood *et al.*, 2001, 5). Others have also argued that the narrow view on technical and statistical risk management seems to be insufficient for the given complexity concerning, for example, government risk strategies and rationalities (Dean, 1999, as cited in Zinn, 2004, 7), emotional and aesthetic (Lash, 2000, as cited in Zinn, 2004, 7) or socio-cultural perceptions and responses to risk. It has also been argued that people do not respond to the same risk in the same way (Tulloch and Lupton, 2003, as cited in Zinn, 2004, 7).

Moreover, Adams argues that the global scale of person-made threats is not a recent phenomenon. He also questions the possibility of maintaining Beck's proposed 'independent' media and court system (Adams, 1995, 185). Other criticisms include Beck's inability to acknowledge the emergence of the concept 'risk' as a specific strategy to manage uncertainty (Zinn, 2004, 6) and that new interests can lead to heightened risk awareness, which in turn can lead to increased political engagement (Zinn, 2004, 7). Beck also fails to take into account the literature of media and communication (Wilkinson, 2001).

*Normal Accidents and High Reliability Organizations*

Perhaps the most popular point of reference for critical infrastructure protection within the field of sociology is the debate in organization studies that is concerned with the safety and reliability of complex technological and social systems. Two schools define the field. High Reliability Organizations (HRO) theory states that hazardous technologies can be safely controlled by complex organizations if the correct design and management techniques are followed, such as strong and persuasive leadership and commitment and adherence to a 'safety culture', including learning from mistakes, creating redundancies and increasing transparency in accountability and operational settings (La Porte, 1996; La Porte and Consolini, 1991; Weik, 1987; Weik and Sutcliffe, 2001). Normal Accidents Theory (NAT), on the other hand, holds that accidents are inevitable in organizations that have social and technical interactive complexity and little slack. According to NAT, the discipline required of an HRO is unrealistic. Systems fail due to their inherent fallibility and the non-responsive nature of bureaucratic organizations. Efforts to increase accountability result in blame-shifting. Indeed, safety is only *one* priority—it competes with many others (Perrow, 1999; Sagan, 1993; Vaughan, 1996).

More recently, growing technical, social and organizational interdependencies refocused the debate from the single organization to networks of organizations. Many scholars now consider *resilience* as the desired objective, which accepts the possibility of massive systems failures due to these complex interdependencies and seeks proactive and reactive strategies to manage a variety of possible consequences (Boin and McConnell, 2007; Clarke, 2005; McCarthy, 2007; Roux-DuFort, 2007; Schulman and Roe, 2007). Within this context, the search for security is seen as a dynamic process that balances mechanisms of control with processes of information search, exchange and feedback in a complex multi-organizational setting, which is guided by public organizations and seeks participation by private and not-for-profit organizations and informed citizenry (Aviram, 2005; Aviram and Tor, 2004; Auerswald, 2006; Comfort, 2002; de Bruijne and van Eten, 2007; Egan, 2007).

*Cultural Theory*

The anthropologist Mary Douglas argues that what a person thinks constitutes risk either to oneself or one's community[7] determines who or what the person blames when things go wrong. This understanding of blame determines the person's accountability system. All three, risk, blame and accountability, are informed by a person's cultural values. This process is a self-reinforcing one more than a chain reaction: cultural values reinforce the accountability system, for example, but the accountability system also reinforces cultural values. The community's institutions, such as the judiciary, uphold this value system. A person's attempt to change these institutions is an effort to argue in support of different cultural values (1992, 24).

Douglas describes a person's value system in terms of the grid/group theory that she developed (Figure 2.3). Grid measures the strength of rules and social norms (1982, 191–2). Group measures the extent to which community constraints are imposed on an individual (1982, 191–2). At the intersection of grid and group, Douglas sees different 'types' of people and community value systems emerging. Each of these different 'types' has different beliefs about what constitutes risk, and what governance structures should be established to mitigate the risk. The central assumption is that there is a relationship between modes of social organization, and responses to risk and culture are adequately represented by the dimensions of the grid/group scheme.

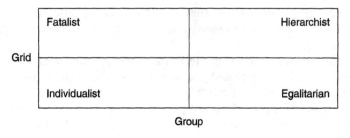

*Figure 2.3*   Cultural theory typology

Based on Douglas and Wildavsky's Grid/Group typology, Hood (1998) and Thompson *et al.* (1990) explore the four 'types' that emerge and the corresponding forms of governance structures that each would develop and their pitfalls.

The hierarchist (high grid/high group) understands good governance to mean a stable environment that supports collective interest and fair process through rule-driven hierarchical organizations. These institutions manage the society's and the individual's resources most effectively: the organization's clearly defined rules and expertise in management enable fair and efficient processes (Hood, 1998, 75). Any departure from this rule-bound hierarchy represents risk for the hierarchist. As such, when things go wrong, hierarchists blame unclear and/or weak rules. Their solution lies in strengthening and clarifying the reporting relationships and the rules that govern the organization (Hood, 1998, 53). Despite the effort to make reporting relationships clear, however, hierarchical systems are susceptible to people working at cross-purposes. Similarly, the vastness of the typical hierarchy allows members to react slowly, absorb significant resources and sweep indiscretions 'under the rug'.

The individualist (low grid/low group) understands good governance to mean minimal rules and interference with free market processes. Individualists believe that people are self-seeking, rational and calculating opportunists. Individual responsibility rules supreme and apathy means consent (Thompson *et al.*, 1990, 34 and 65). In contrast, individualists understand risk to be government regulation of the economy or the management of public services. Despite the individualist's faith in market practices, individualist practices have their own pitfalls. Within an organization, individualist practices, such as pay-for-performance, can undermine collective goals and lead to a lack of cooperation as employees compete for salary increases. Moreover defining the public servant/ citizen relationship as 'customer/producer', as individualists are wont to do, changes the nature of the relationship and leaves some more vulnerable than others (Fountain, 1999, 2).

The Egalitarian (high group/low grid) understands good governance to mean local, communitarian and participative organizations. For egalitarians, authority resides with the collectivity. Moreover, organizations are flat, or at least there is minimal difference between top official and rank and file. Fellow workers, not superiors, conduct performance appraisals. And in order to maximize transparency, maximum information is available to workers and to the public. Egalitarians understand risk to mean hierarchies and organizations outside their system. When things go wrong, Egalitarians blame externals: 'management', 'the executives', 'the system' and 'Wall Street' (Thompson *et al.*, 1990). Egalitarian systems strive for equality, but frequently miss the differences. Egalitarian organizations are susceptible to treating everyone in the same manner. This one-size-fits-all egalitarian approach can often result in splits or breakdowns in the organization as individuals strive to define themselves.

The fatalist (low group/high grid) understands good governance to mean management by surprise techniques, or by circumventing practised or routine responses. Good governance anticipates lack of cooperation between citizens in a chaotic and unpredictable universe (Thompson *et al.*, 1990, 35). The randomness that makes up fatalist forms of governance undermines incentives to innovate, develop or compete. While one of the benefits of 'contrived randomness' is that it might prevent collusion by randomly reassigning employees to different teams, or parts of the organization, such randomness would also undermine the incentive to build-up strong teams.

Cultural Theory has had limited success when tested empirically. (See, for example, Dake, 1991; Sjoberg, 1997.) Dake had some success but noted the correlations between culture and bias were weak and of limited predictive value. The grid/group typology is also criticized on the grounds that the categories in the typology are too limiting. Assumptions about risk perception are far more complex and dynamic than the categories imply (Renn *et al.*, 1992) and Cultural Theory also fails to take the media into account (Zinn, 2004, 15).

### Some Interim Observations on the Risk Literature as Mapped on the Risk Rationality Diagram

Let's note some broader trends and insights that the Risk Rationality diagram reveals. First, due to the different understandings of the source and nature of risk, the solutions to risk problems differ according to each type. In the lower half of the diagram, in which risk is an objective reality, solutions lie largely in 'design'. For those in the Individual/Objective section, for instance, problems can be understood and the solutions to those problems lie in understanding and fixing the problems based on their technical characteristics. For those in the Structural/Objective section, the risks can be understood in a similar way but the problems and solutions go beyond the individual; they are embedded in the structural setting in which the problems exist. Therefore, the solutions lie in institutional design. While Habermas was not especially hopeful that the administrative systems that perpetuate inequality could be overcome, one might still suggest that those who hold views in the lower half of the diagram are much more optimistic. Problems can be identified. There is a clearer relationship between cause and effect. Context is not a determining factor here: if different actors have access to the same information and they all act 'reasonably', all will arrive at the same conclusions.[8]

On the top half of the graph, where approaches to risk are still considered rational (albeit a bounded rationality) but are either individually or socially constructed, there is no 'one-view-fits-all' understanding to risk. For the Individual/Constructionists, the solutions necessarily lie in appealing to individual perceptions, and therefore persuasion and the manipulation of symbols play an intricate part of any risk solutions. For the Constructionist/Structural type, cause and effect relationships are context-dependent. The understanding of risk varies with the institutional setting in which members exist, and therefore the solutions vary accordingly. Academics working in this half of the schema tend to be much more pessimistic about finding stable solutions to risk-related problems than those working in the lower half. Critical concepts such as 'democracy', 'fairness', 'equality', 'transparency' and so forth will always be debated. Indeed, even the decision to opt for a technical or design solution is viewed through a socio-political lens.

Finally, there are those in the middle—the advocates of Systems Theory, Risk Society and Normal Accidents Theory. While these theories may vary somewhat on the question of whether or not risk is a constructed concept, the inevitability of risk and uncertainty in today's highly complex modern society, which depends on sophisticated and interdependent societies and technologies, joins the three. HRO's optimism is an outlier here. Cause and effect relationships are in flux. Proposed solutions offer faint hope. Problems and their solutions are more contentious and contested; the solutions are always provisional. The world is complex, particularly in the structural side. In this section,

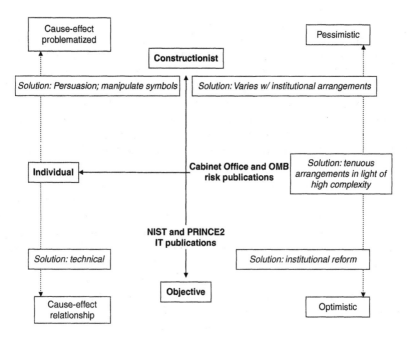

*Figure 2.4* Risk Rationality Diagram, amended

Source: Based on the diagram presented by Renn, at the Social Contexts and Responses to Risk Inaugural Conference, January 2005, at the University of Kent at Canterbury.

solutions work towards achieving stability in a complex, contested environment. The amended diagram above elucidates these points (Figure 2.4). It also includes where on the schema one would map the key government publications on risk and IT management.

While the schema does help to generalize on some important starting points, caveats about the diagram should be noted.

The diagram represents two *continua* and should not dictate 'either/or' thinking. For instance, the individual/structural continuum is challenged by approaches to risk that seem to fuse personal perceptions with social movements. Kasperson's (1992) Social Amplification of Risk and Cohen's (1972) study of Moral Panics are models that seem not simply to exist between the two extremes of the individual and structural but rather seem to fuse the two concepts. Kasperson and Cohen describe a dynamic process that depends simultaneously on individual perceptions, social amplifiers and institutionalized powers that influence jointly which risks will predominate.

Moreover, many practitioners and academics would claim to be in the Individual/Constructionist section or the Objective/Structural section when in fact they are merely RAP in disguise. Decision analysis techniques and cost benefit analyses could be categorized as individual/constructed but so too could they be described in the same vein as probabilistic risk assessments, as individual/objective. Perhaps this point takes its inspiration from Critical Theory and relates more to a normative point about process and the question of how these tools are applied. While Jaeger *et al.* (2001) might describe decision analyses as a form of creating consistency in decision-making while keeping the 'constructionist' component intact, decision analyses are not necessarily pitched that way to users. There is not necessarily an effort to expose the underlying assumptions in these tools. Data, presentation and scales and lay people's 'faith in numbers' (among other things) are ripe for manipulation by analysts. The same can be said of these tools in organizational settings. The importance of organizational context or individual preferences can merely be a footnote for many academics and practitioners who employ such tools. In short, if the lessons of Critical Theory, which expose such limitations, are not pursued, these tools will continue to experience the limitations of RAP in practice if not in theory.

Finally, with respect to government trends in IT and risk management, the Risk rationality diagram helps to elucidate a gap that emerged between the communications and policy direction advocated by central agencies about how to manage 'risk' in generalized terms and how central agencies as well as departments and agencies are pursuing risk management of IT in particular. This contradiction is not new and in fact as this research will show it existed throughout both governments' approaches to Y2K. IT units applied RAP-based tools to a problem that was perceived as pervasive and unknowable. It necessarily led to a massive reaction, the size, structure and style of which the next chapter aims to explore.

# 3
# How Did the Governments React to Y2K?

Both governments articulated an ambitious goal for the Y2K operations within their respective government departments and agencies—there was to be no disruption to service. The bureaucratic processes that supported the goal were largely the same in both countries: top-down, slow, detailed, expansive, exhaustive, template-driven and resource-intensive. There were also significant overlaps in the form of mandatory third party audits and contingency plans. Each government's interactions with industry on the issue were similarly expansive: both governments reached out to a large swath of industry and encouraged information-sharing and standard-setting *within* but also *across* all the critical sectors of the economy.

While the Y2K operations in both countries may have ended this way, however, neither started this way. None of the agencies at the operational level, initially, took the problem as seriously as the oversight offices eventually did. Many were loathe to break from their regular work routines. This resulted in false starts, inconsistent reporting and delayed progress. As 'outsiders' began to ask questions and to infringe on departmental space—with prescriptive orders, timelines, templates, audits—front-line staff became aggravated. Y2K work was often boring, detailed and tedious, supplemented by (sometimes) weekly detailed, reporting requirements. In some cases staff were confident either there were no Y2K-related problems or they could fix them as part of their normal process, yet these observations went unheeded, even by the executives within their own agencies, which led to further confrontation. Like it or not, staff were corralled into large Y2K operations where risk *management* meant risk *elimination*.

Yet while the grasp was significant, the reach was oftentimes less so. Despite the novelty of the problem, their approaches followed long-standing institutional practices and arrangements, in which significant tensions and trade-offs are embedded. No solution was perfect; each had its compromises, both within government and outwith.

This chapter describes and analyses comparatively the US and UK governments' management of Y2K according to the Hood *et al.*'s categories of Size, Structure and Style of management.[1] The discussion is aggregated at the agency level (that is, the executive, the statistics agencies and the aviation agencies). I grouped the agencies in this way in order to bring the comparative dimension into focus more acutely. We will start with the Executives' orders and will then turn to the agencies' responses.

## The size of the Executive's response

Size can be conceived of in two separate ways: (1) *Aggression,* the extent of risk toleration in standards and behaviour modification, and how far regulators go in collecting information about the risk; and (2) *Investment,* how much goes into the regime from all sources (Hood *et al.*, 2001, 31).

Both executives set tough standards for their respective departments and agencies. Following a speech on the topic by Prime Minister Blair (1998), the Cabinet Office (CO) promised 'no material disruption to essential public services upon which the public rely' and enforced that goal by requiring that departments and agencies complete and submit to the CO detailed, standardized templates on Y2K progress. Similarly, President Clinton signed Executive Order (EO) 13073, which promised 'no critical Federal programme would experience disruption because of the Y2K problem' (cited in President's Council on Year 2000 Conversion, 2000, 22). The President also addressed the issue publicly but not until four months after Blair did (Clinton, 1998). Similar reporting templates were issued in the United States and the United Kingdom. While at first glance both governments' template-reporting within departments and agencies may seem like a form of information gathering, by early 1998, after Blair's speech and EO 13073, the templates were in fact forms of standard setting and behaviour modification. The templates came not to mean 'Are you Y2K compliant?' but rather '*When* will you become Y2K compliant?' Virtually all systems would follow a standard, guaranteed-to-work approach: inventory, fix, test and audit every system[2]. Reporting requirements were quarterly, then monthly. Summary reports were made public and there was low tolerance from any oversight body for anything less than a form of demonstrable or verifiable Y2K compliance, which meant following the standard process.

Prior to the President's intervention the Congress and the General Accounting Office[3] had been relentless in applying pressure and scrutiny on the Executive. Between 1996 and 1999 congressional committees and subcommittees held over 100 hearings on Y2K and the GAO issued 160 Y2K reports and testimonials. The House of Commons was more muted. While the NAO played an important role in scrutinizing government progress on Y2K, it published only seven reports and the Public Accounts Committee held only three hearings on the subject.

Despite the pressure from the respective legislatures and the high profile interventions by the Prime Minister and the President, OMB and CO had already been pushing the Y2K issue forward for some time. The CO[4] had already directed each department and agency to audit its systems by January 1997; prioritize and cost a programme of action by October 1997; and test all modified systems by January 1999, except for financial systems, which were given a deadline of April 1999 (NAO, 1997, 6). Initially OMB targeted 24 major government departments and their related agencies (1997b). Originally they were expected to renovate their systems by December 1998 and implement all changes by November 1999. Concerns over departments' falling behind, however, resulted in deadlines being advanced. By December 1997, OMB accelerated the deadlines by three months (September 1998) and eight months (March 1999), respectively. By March 1998 OMB also expanded its scope and directed all agencies—big and small—to submit their Y2K plans to OMB quarterly (Table 3.1). Hence, ultimately, both executives had similar scope, standard and timelines for their departments and agencies.

Oversight agencies learned important points as they progressed. For example, Y2K was not simply about having one's system ready on 1 January 2000, but rather ensuring that one's system could process a year 2000 entry whenever it needed to. As a result many systems had to be ready (at the very least) by the start of the fiscal year 1999/2000, which of course for many meant 1 April 1999. Equally important, Y2K operations progressed more slowly than originally planned and therefore some slack had to be built into the process. Finally, it was expanded to include all departments and agencies because there continued to be considerable uncertainty about the magnitude of the problem and agencies' capacity to withstand it. As we will see, time and time again, oversight agencies expanded reporting requirements in an effort to get a fuller picture, which often remained elusive.

*Table 3.1* Cabinet Office and Office of Management and Budget Directives to Government Departments. CO directives were sent to all departments and agencies. OMB directives included departments only, and then expanded to include all departments and agencies

| UK | | US | | |
|---|---|---|---|---|
| CO Directive | Deadline | OMB Directive | Original deadline | Revised deadline |
| Inventory | January 1997 | Renovate | December 1998 | September 1998 |
| Prioritize and Cost | October 1997 | Implement all changes | November 1999 | March 1999 |
| Test all modified systems | January 1999, (financial systems, April 1999) | | | |

The approach across the infrastructure was also expansive. By way of promoting Y2K compliance across the entire infrastructure, both governments identified key sectors as making-up the national infrastructure and created voluntary fora in which representatives from each of these sectors could share information on Y2K and report on their sectors' degree of readiness as the date-change approached. The UK Government grouped 25 sectors in its Y2K National Infrastructure Forum (UK/NIF) and the US government identified 26 sectors in its Y2K Working Groups (US/WG). The standards for compliance were established at the sector-level. The Senate, the NAO and both governments' lead offices on Y2K (the US Government's President's Council on Year 2000 Conversion and the UK Government's Action, 2000) all issued summary reports throughout 1999 on the Y2K status of each of the sectors they had identified. Table 3.2 below lists the sectors represented in the UK/NIF and the US/WG. The sectors in the UK/NIF are organized by tranche.[5] In the United States, the sectors are listed in alphabetical order.

The perceived dependence on complex national and global supply chains, and in particular the role of small and medium-sized enterprises (SMEs) in those supply chains, however, meant that traditional sectors were not the only ones whose Y2K compliance was relevant to the functioning of the economy. Both governments' Y2K strategies also included informing harder-to-get-at groups about Y2K. In order to reach these audiences both governments put more emphasis on communications, particularly in 1999. The Media Communications Unit (MCU) at the CO began

*Table 3.2* Critical sectors represented at the National Infrastructure Forum (UK/NIF) and the Working Groups (US/WG)

| Tranche | National Infrastructure Forum – UK (25) | Working Groups – US (26) |
|---|---|---|
| 1 | Electricity | Benefits payments |
| | Gas | Building and housing |
| | Fuel supplies | Consumer affairs |
| | Telecommunications | Defence and international security |
| | Water and sewerage | Education |
| | Financial services | Emergency services |
| 2 | Essential food and groceries | Employment-related protections |
| | Rail transport | Energy (electric power) |
| | Air transport | Energy (oil and gas) |
| | Road transport (Local | Financial services |
| | government) | Food supply |
| | Sea transport | Health care |
| | Hospitals and health care | Human services |
| | Fire service | Information technology |
| | Police | International relations |
| | Broadcasting | International trade |
| | Local government | |
| 3 | Sea rescue | Non-profit organizations and |
| | Weather forecasting | civic preparedness |
| | Post and parcels | Police and public safety |
| | Welfare payments | Small business |
| | Flood defence | State and local government |
| | Criminal justice | Telecommunications |
| | Tax collection | Transportation |
| | Bus transport | Tribal government |
| | Newspapers | Waste management |
| | | Water utilities |
| | | Workforce issues |

*Notes*: The UK/NIF sectors are divided by tranche. The US/WG sectors are listed in alphabetical order.
*Source*: Action 2000, 1999; President's Council on Year 2000 Conversion, 2000.

side-stepping the media and paying for means of communicating with the public directly. They held public events, paid for advertising in papers and issued numerous publications to various audiences, including SMEs and micro-businesses (MBs) (Cm 4703, 2000, 32).[6] In addition, Action 2000 collected and published survey data on SMEs and MBs. In the United States, the President's Council organized Y2K Action Weeks (which focused on small businesses), published resource guides (for various audiences) and organized 'Community Conversations' (that is, town hall meetings) to answer the questions of ordinary citizens (President's Council on Year 2000 Conversion, 2000, 7–12).

The difference in cost between each government's respective Y2K programmes is striking and seems to go beyond the fact that the US government was more dependent on IT than the UK government was. Ultimately, the UK government declared its Y2K operation cost £400 m. Expensive? Yes, but the fact that each department funded its own plan from its existing IT budget almost certainly acted as a constraint on expenditures. In short, there was no 'new' money for Y2K operations in the United Kingdom. In contrast, in the US government, Y2K funding came almost exclusively from a special appropriation, which OMB administered and most departments saw as entirely 'new' money. Clearly, there were incentives to generate Y2K proposals, and big ones: the US government spent $8 billion in total—13 times more than the government of the United Kingdom.

The US Commerce Department estimated that businesses and government in the United States spent about $100 billion between 1995 and 2001 on Y2K (President's Council on Year 2000 Conversion, 2000, 13). Action 2000 estimated that the UK government, including local authorities and the NHS, spent approximately £1 billion (Cm 4703, 2000, 76). The limitations of these estimates and determining Y2K spending in general will be addressed in the Market Failure Hypothesis chapter.

## Reporting structures

Structure overlaps with size to some extent. It refers to the way that regulation is organized; what institutional arrangements are adopted; and the way resources invested in regulation are distributed. Structure can be conceived of in at least two separate ways: (1) the extent to which regulation involves a *mix* of public and private sector actors, including the use of intermediaries; and (2) how *densely* populated the regulatory space is by separate institutions, and how far the risk involves multiple, overlapping systems of regulation (Hood *et al.*, 2001, 31).

Prior to the two governments' setting their standards for Y2K compliance, there were not that many structures in place specific to Y2K. Initially, Y2K monitoring was built largely on existing structures: OMB/CO; legislative committees and auditors; and departmental executive committees, with perhaps the only notable exceptions being the United States' Interagency Year 2000 Committee, which coordinated government agency activity on Y2K, and the United Kingdom's Taskforce 2000 (Department of Trade and Industry (DTI, 1996), which was initially set up to raise the profile of Y2K across all sectors but ultimately was considered ineffective and was replaced by Action 2000.

The absence of Y2K-specific structures had marked strengths and weaknesses in the United States in particular. On the other hand, there was not really one focal point at which the Y2K debate occurred; information was very piecemeal. Indeed, Y2K hearings were occurring at almost every congressional committee. On the other hand, this dynamic created in effect a Y2K chorus right across Capital Hill that demanded more effective action from the Executive on Y2K.

Political intervention from the Executive Office marked a seismic shift. Only when the political levels intervened in late 1997 and early 1998 did the governments create new committees within the executives, such as MISC 4 (Y2K Cabinet Committee), the President's Council on Year 2000 Conversion, Action 2000 and the voluntary sector level groups noted above, the National Infrastructure Forum (UK/NIF) and the Working Groups (US/WG). There was also a corresponding growth in the Senate with the creation of the Special Committee on the Year 2000 Technology Problem, which mirrored the creation of the President's Council on Year 2000 Conversion in the Executive Office of the President (EOP). The new structures were largely devised to change behaviour—to get people and organizations *moving* on Y2K strategies. There was a high-level of awareness about Y2K but uncertainty remained about outcome, which these organizations aimed to fix.

Because of the task the governments wanted to accomplish—to ensure Y2K compliance across the entire national infrastructures—the project necessarily involved the participation of both public and private sector actors.[7] The US approach assumed a pluralist mould with private industry; the United Kingdom assumed a corporatist one. In many respects these interactions replicate the manner in which the two governments normally interact with industry (Vogel, 1986). Membership of the US/WGs included major industry trade associations and other umbrella organizations representing the individual entities operating in each sector. The UK/NIF in contrast included a small group of influential organizations within each sector that represented a significant share of the industry. Appendix C lists as an example the members of the Transportation Working Group and the UK/NIF's Aviation Group. Clearly both governments were trying to accomplish two competing goals: (1) to share information with as much of the industry as possible (however defined); and (2) to keep these groups to a manageable number. The UK/NIF's Aviation Group drew from 15 organizations. The US/WG on Transportation drew from seven government departments and 22 associations.

The difference between the governments' approaches is significant. In the United Kingdom, the individual organizations, for example, British Airways, were members of the UK/NIF and therefore shared some responsibility as well as helped to create and endure the group pressure that could potentially be exerted on the UK/NIF. In the United States, in contrast, intermediaries acted as a buffer on behalf of industry. And that was not simply in the US/WGs; it was also the case at congressional testimonies. US congressional committee staff members noted that trade industry representatives appeared before committees and in so doing buffered specific organizations and individuals from answering difficult Y2K-related questions posed by committee members (INT 39). Indeed, it was difficult to get corporate executives to agree to appear before Senate committees to provide public evidence on their organizations' Y2K readiness.

It was not only the government that was attempting to tread onto the private sector's turf, however. The reverse was also true. The private sector weighed into government Y2K operations in two ways. First, private industry was equally concerned that the government as a key provider of numerous public services would also be ready in time for Y2K. Second, given the scope of the task that faced both governments with their own systems, they had to look to the private sector for help. Both governments depended on private sector counsel to fix and verify the governments' Y2K problems. From OMB/CO down to departments, the private sector played a large part in advising, fixing and auditing the government's bug problems.

Figures 3.1 and 3.2 summarize the reporting structures in both governments. Representatives from the private sector are strongly present in both diagrams. In the United Kingdom, one can see private sector presence at Number 10, the CO and the UK/NIF as well as among those working at the department/agency level. In the United States, private sector presence comes mainly in the form of the US/WG, and Special Advisors Group (SAG) as well as among those at the department/agency level; however, those in the US/WG were largely representatives from industry associations. Only the members of SAG were leaders from specific organizations. Typical of the pluralist/corporatist distinction, the private sector representatives in the United States seem to be less deliberately integrated at the executive level than those in the United Kingdom. Note, also, that Congress, which

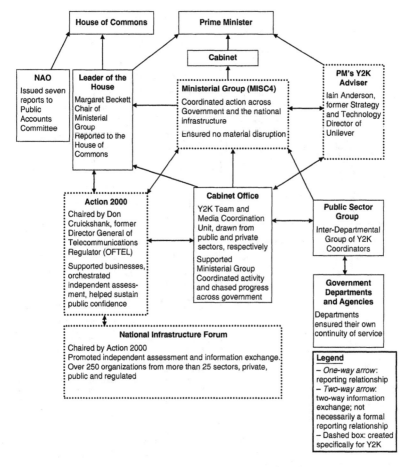

*Figure 3.1*   Final UK government Year 2000 reporting structures
*Source*: NAO, 1999a, 13; Cm 4703, 2000, 69.

was also a 'joining-up' point between government and industry, played a much stronger role than did the House of Commons in the United Kingdom.

## Style of the government's response

Style overlaps with the other two descriptions of management. It can be conceived of in two ways: (1) how far regulation is *rule-bound or discretionary*; and (2) the degree of *zeal* the actors show in pursuit

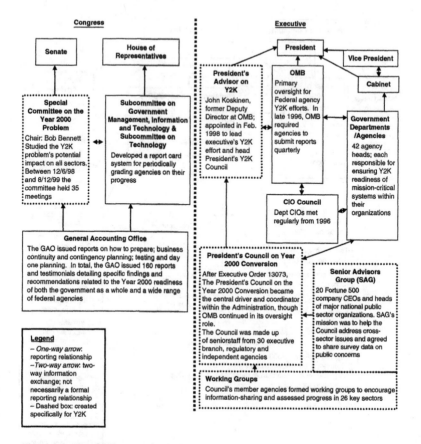

*Figure 3.2*   Final US government Year 2000 reporting structures
*Source*: Special Committee on the Year 2000 Technology Problem, 2000; President's Council on Year 2000 Conversion, 2000.

of policy objectives. With style, culture and attitudes are important (Hood *et al.*, 2001, 32).

In some respects the response to Y2K within departments and agencies has Weberian zeal to it—massive, command and control, detailed, thorough, slow, resource-intensive, inflexible and process-focused. They were devised as top-down approaches; the only feedback loops that were put into the process were those that confirmed that departments and agencies were progressing towards compliance as defined by central agencies. Indeed, the OMB Y2K process deliberately mimicked its budgetary approval process. Ironically, agencies may not have found many Y2K-related problems but the governments did not construct any way of gathering that information.

Despite the seemingly vast and thorough approach, there were loopholes and gaps. While the governments' Y2K paper-trails were massive and deadlines were fixed, many government departments (and UK/NIF participants) still missed government deadlines regularly. Indeed about 25 per cent of UK government bodies did not finish Y2K work until the last quarter of 1999. (See, for example, NAO, 1998; 1999a; and OMB, 1997a,b,c; 1998a,b; 1999a,b,c). Agency and executive staff often acknowledged that deadlines were rather artificial because both the executive and the front lines knew that they would not be met. And while the NAO insisted that Y2K publications were approved for publication much more quickly than other reports (INT 60) by the time the reports were published they were often weeks out of date due to the various checks and sign-offs that were required as the reports made their way up the reporting chain. This slowness had its advantages. It gave departmental staff lead-time to fix any problems for which they would be criticized in publications before the documents were made public.

Though departments and agencies tended to be the slowest, the entire UK/NIF experienced difficulties in meeting timelines. Figure 3.3, for example, summarizes the progress of the UK/NIF in 1999 as the sectors inched towards verifiable compliance. Note, the UK/NIF included many government departments and agencies that were supposed to be finished by 1 April 1999 at the latest. Indeed, neither government could claim their systems were compliant until the very end of the year (the United Kingdom in November and the United States in December).

Moreover, it is unclear just how air-tight the Y2K plans were. While agencies from both countries' departments sent letters to their respective supply chains seeking Y2K assurances, the NAO found that in the United Kingdom response rates from suppliers ranged from '100%' in some departments to '0' in others (NAO, 1999a, 36). Equally, external auditors were loath to give assurances that agencies were compliant because it was difficult to be certain; they opted to audit the *process* only. These gaps did little to reassure the governments, and in fact resulted in the central agencies demanding more contingency plans to guard against unforeseen failures.

Institutional inertia provoked reactions among onlookers, however, particularly in the United States. Information-gathering and behaviour modification in the United States could be more personally intrusive than in the United Kingdom. In some cases Vice President Gore confronted

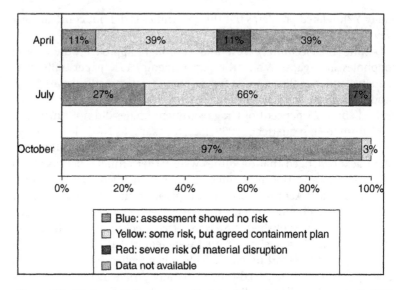

*Figure 3.3* National Infrastructure Forum: subsector progress towards Y2K Compliance, 1999

*Note*: The 25 sectors were subdivided into 61 subsectors. Any subsectors showing any degree of 'red' were placed in the red category. Subsectors showing no red but at least some yellow were placed in the yellow category. Only those subsectors that were graded with a 100 per cent blue rating are captured under blue.

agency heads personally when their agencies fell behind (President's Council on Year 2000 Conversion, 2000, 5). Congress and the GAO grilled agency heads publicly on Y2K. The House of Representative's Subcommittee on Government Management, Information and Technology reduced agencies to a letter grade (GMIT, 1999) with the intention of embarrassing slow agencies into action, as one committee staff member noted (INT 39).

What is perhaps most striking is the unusual level of collaboration within both countries between the executives and the legislatures in developing and policing the government strategy among agencies. The CO and the NAO and the OMB and the GAO, respectively, worked together, formally and informally, sometimes sharing the same office space and the same source documents.[8]

From the organizations in the infrastructures, both governments gathered information only at the sector level; these fora were voluntary and therefore tenuous; nobody was formally obligated to be there. While the US Government passed legislation on Y2K and the United Kingdom did not, *The Year 2000 Information and Readiness Disclosure Act*

(October 1998) and *The Year 2000 Readiness and Responsibility Act* (also called *The Y2K Act*) (July 1999) were enacted to create an environment that was similar to that of the United Kingdom where information could be shared among organizations with less fear of being sued if the information turned out to be incorrect.

Nevertheless, there were some significant differences in each government's approach to the infrastructure. EO 13073 directed the government to help industry but it was up to industry to sort its own problems out. Members of the UK/NIF, on the other hand, worked under a variation of one government strategy, 'no material disruption', rephrased for the UK/NIF to 'Business as Usual'. But members of the UK/NIF represented a consensual arrangement among a small sample of organizations from each sector. The NAO and Action 2000 therefore made generalized claims about the sectors, even when for reasons of complexity and volume such claims were contestable, if not doubtful. Witness one organization dropping out of the UK/NIF in the aviation sector, for instance, yet the government still reported the sector as being Y2K compliant (NAO, 1999b and Action, 2000, 1999). The UK/NIF was not just about the technical nature of the risk; there was also a public relations dimension to it. Ultimately, each sector was *going* to be declared Y2K compliant.

## Management at the departmental level

### Size of the response

#### Statistics agencies

All four agency case studies are almost entirely dependent upon technology to deliver their respective service. Their task was daunting. Ultimately the statistics agencies had detailed and thorough responses to Y2K. Although there was a slowdown in the case of the ONS and a slow start in the case of the BLS, they eventually followed the central directives: inventory, fix, test and audit. It was a slow, step-by-step process that required considerable time and resources.

The ONS started early and implemented a programme with considerable scope but like many agencies soon lost momentum. The ONS launched its Y2K programme in the summer of 1996. It established a Y2K programme board and appointed a (part time) manager to coordinate Y2K activity and collect reports from business areas. The initial phase of the programme concentrated on developing an inventory of applications, software and infrastructure; raising awareness of Y2K throughout the ONS; and developing roles, responsibilities and timelines (ONS, 2000, 4). After what was described by an external auditor as a 'good start', ONS progress slowed. In January 1998, external auditor Impact concluded there had been considerable slippage from the original timelines and the ONS was in jeopardy of missing government deadlines. The ONS briefing material noted the task was bigger than it had originally thought.

The BLS in contrast was slow off the starting blocks, and soon faced growing pressure from oversight agencies. IT staff within the BLS had been aware of the Y2K problem for years but failed to make marked progress. Indeed, one director noted that the first time the BLS had to deal with a Y2K problem was in 1985—to fix the Employment Projection System. He recalls issuing a memo on the potential consequences of the date-related problem in the early 1990s but it seemed to have had little effect (INT 27). The lack of impact was not lost on Congress: in July 1996 in its first Year 2000 Progress Report Card, the Subcommittee on Government Management, Information and Technology (GMIT) issued an 'F' for the parent department, the Department of Labor (DOL), one of only four agencies (out of 24) to receive the lowest grade. By May 1997, almost a year after the ONS had started its Y2K programme board, the DOL directed each of its agencies, including the BLS, to appoint a Y2K Project Manager. From May 1997, the DOL's CIO met monthly with its agencies' Year 2000 Project Managers and IT Managers (GAO, 1998d, 4). Despite these initial steps and the Secretary of Labor declaring in a December 1997 memo that Y2K was a departmental priority, reports from GAO and OMB continued to show slow progress at the DOL and its agencies. By August 1998, for instance, only 11 out of 23 (48 per cent) of the BLS mission critical systems were considered compliant. (See, for example, OMB, 1997c, 1998a, 1999b; GAO, 1998d.)

Both agencies improved their performance from the central agencies' perspectives in 1999. Following an external auditor's report, the ONS engaged a full-time manager and increased the staff size in the Y2K unit. Y2K Programme Board meetings increased to monthly, as did the reports to the ONS Board. In addition, the Y2K project staff developed explicit and detailed logs that tracked the status of every system in the organization. Nevertheless, a second audit revealed that some 'high criticality' systems were not scheduled for completion until September 1999, well after the CO's deadlines. The BLS, for its part, also moved towards compliance much more rapidly (GAO, 1999b). By March, OMB reported that the DOL had indeed made good progress on mission critical systems—from 67 per cent compliant in November 1998 to 85 per cent compliant (OMB, 1999a).

As the project expanded so did the budgets. The ONS initially forecast £2.47 million for its Y2K programme. The initial forecast was revised to £2.67 million in early 1998 and then to £2.9 million in early 1999. Ultimately the ONS spent £3 million on its Y2K programme, an increase of 21.5 per cent over the original forecast. In total the DOL estimated it would spend $60.4 million on Y2K (OMB, 1999c, 53). Approximately 50 per cent of this total was spent in 1999.

*Aviation agencies*

No government agency was under more pressure than the FAA, largely because of its late start, but ironically it was the CAA that was progressing the more slowly of the two. Like the statistics agencies, the aviation agencies followed the standard approach: inventory, fix, test and audit, with few exceptions. But that seemingly exhaustive approach was only part of the challenge: the regulators were ostensibly accountable for a vast and sprawling aviation industry, also heavily dependent on technology. Here, their efforts diverted.

In January 1998 in response to a request made by Congress to audit the FAA's Y2K plan, the GAO reported: the FAA had no central Y2K programme management; an incomplete inventory of mission critical systems; no overall strategy for renovating, validating and implementing mission critical systems; and no milestone dates or schedules (GAO, 1998b). In addition, until the aforementioned had been accomplished, the FAA's cost estimate $246 million was unreliable. More importantly, the report concluded that 'should the pace at which the FAA addresses its Year 2000 issues not quicken...several essential areas—including monitoring and controlling air traffic—could be severely compromised' (GAO, 1998b, 15). Following the GAO report and the FAA's appearance before the congressional subcommittee on Government Management, Information and Technology,[9] the FAA started its Y2K project *de rigueur*. The FAA established (1) an overall strategy; (2) detailed standards and guidance for renovating, validating and implementing mission critical systems; (3) a database of schedules and milestones for these activities; and (4) a Y2K business continuity plan (GAO, 1998c). According to every interview subject, money was not an object. The person that coordinated the FAA financial requests simply had to call the OMB and request the money. The FAA was never refused. One middle manager noted how, on one occasion, he signed personally for a $15 million upgrade during the 1998/1999 period, something that would have been unthinkable and organizationally unjustifiable a few months earlier (INT 44).

The CAA started its Y2K programme at approximately the same time as the FAA. Like most organizations, the CAA started by conducting a series of scoping studies to determine what inventory it had and which of the systems could be replaced (as opposed to renovated) before 2000. The scoping was completed by the summer of 1998. The CAA grouped its internal computer systems according to their importance (as determined jointly by programme areas and the IT department): *level one* (business critical); *level two* (business important); *level three* (other supported applications); and *level four* (other supported applications of lower importance).

Briefing material notes that most changes were cosmetic only, though often not completed until October 1999, including several level one and level three systems. The National Air Traffic Service (NATS) fared better. It finished its initial round of testing by the summer of 1998, though some systems required additional work. In March 1999, NATS declared it would be 'business as usual' over the millennium (NATS, 1999b). In the main, the FAA's approach, like that of the CAA, was to renovate the system. Occasionally it replaced systems, particularly if the systems were scheduled to be replaced shortly after 2000. The FAA completed work on 424 mission critical systems and 204 non-mission critical systems and was declared compliant by June 1999, less than 18 months after the GAO's highly critical report. The work had been examined by Science Applications International Corporation (SAIC), an independent verification and validation contractor. The Office of the Inspector General (OIG) from the Department of Transportation had also examined a sample of systems and approved the work.

With respect to the aviation industry the CAA had two Y2K programmes in place: *The Safety Programme* and the *Business as Usual Programme*. In effect, the Safety Programme reflected normal CAA business practices. It involved issuing memos to the industry on the topic of Y2K in 1997 and 1998, and in 1999 beginning an assessment of the industry's readiness. The CAA expected Y2K accomplishment statements from all of the 1880 organizations that it regulated. It also conducted audits of between 8 and 10 per cent of the organizations.

Following Number 10's intervention and the creation of the UK/NIF, however, the CAA supplemented *the Safety Programme* with the *Business as Usual Programme*. Business as Usual was a voluntary programme that produced sector-wide summaries, which initially involved the top 15 companies in the UK aviation industry.[10] As part of the *Business as Usual* programme, the CAA contracted AEA Technology Consulting to conduct independent surveys of the participants. AEA Technology reviewed questionnaires, conducted site visits and developed short assessments in consultation with the participant organizations that were based on Action 2000 definitions of readiness. In particular AEA examined management structures, sampled documents, checked strategies and processes, identified risks and ensured reasonable business continuity plans were in place. Sector-wide results were then published.

For the FAA, there was essentially only the standard practice vis-à-vis the aviation industry, with some exceptions. Interview subjects were adamant that they could not guarantee that the entire aviation industry would be bug free—it was just not possible, operationally. Rather, the FAA did what it always did; it conducted audits of a certain percentage of the industry, but added a few Y2K-related questions and sent information on the topic to all licence holders. While the FAA was confident that the aviation industry would be safe in the cross-over to the New Year, it did not speak on behalf of the aviation industry (INT 45).

Y2K financial estimates are almost always rough and broad. The Air Transport Association estimated that internationally airlines spent $2.3 billion on Y2K. BA, for example, the largest service provider in the United Kingdom, spent £100 million (Bray, 1999). The US Department of Transportation, of which the FAA's Y2K costs were a part, spent $345.8 million on Y2K between 1996 and 2000 (OMB, 1999c, 51). I did not find a record of how much the CAA spent.

### Reporting structures

*Statistics agencies*

Oversight bodies penetrated agency work routines more and more as the Y2K programme continued. Both agencies had central Y2K reporting offices but they both acted as conduits. Primary responsibility was decentralized to programme areas which, in effect, were working according to CO and OMB directives, which were reinforced frequently by detailed reporting requirement and process audits.

Both agencies had central Y2K boards and units that were tasked with coordinating Y2K efforts in what was a relatively decentralized environment operationally. The ONS collected Y2K information centrally, which it collated, summarized and sent to the CO, but the ONS made business areas responsible for compliance, both operationally and fiscally. The BLS was even less centralized at the agency level. The BLS also had a Y2K operational committee but it met infrequently. Most direction was issued by e-mail directly from the DOL and with limited scope for input from the BLS. Therefore, the BLS Y2K staff (one director and one analyst) simply forwarded directions to individual programme areas. When Y2K

IT staff received the results from the programme areas, they collated the reports, summarized them and submitted them to the DOL.

There was considerable overlap—or at least institutionalized 'double-checking'—within Y2K projects in the form of audit. Both agencies were subject to internal and external audits. The BLS established and operated its own internal verification and validation process (IVV). The BLS felt it was sufficiently independent of the operational units of the agency to provide an adequate check of Y2K compliance (INT 27; 41). In addition to the IVV, the DOL's Office of the Inspector General (OIG)[11] audited the BLS three times. The OIG contracted external auditors to carry out its audits.

The audits detected weaknesses in the reporting structures, and in particular the ONS's system-by-system approach, rather than a process-by-process one. The audits cited many related shortcomings: inability to define 'a system' (that is, when does a 'system' begin and end); lack of intra/inter business area coordination; and inability to assign a system owner to take responsibility for outcomes. Prior to the March 1998 audit by Impact, most of the Board members came from the IT community. After March 1998, the project structure was reassessed. The reassessment resulted in each business area having a Y2K project board. Despite these efforts, however, an audit conducted later by KPMG (June 1999) drew the same conclusions, that contingency plans were systems based, not process-based, which continued to produce gaps in accountability.

*Aviation agencies*

Internally, the FAA and the CAA had similar operations: each had a central Y2K office, first operational in 1998, collected data and reported up the chain to OMB/CO. Both also relied on external organizations to help them reach Y2K compliance—through audits and project support. Both aviation regulators, structurally, were forced to integrate with industry in the form of special Y2K committees, such as the UK/NIF and the US/WG. The vast and sprawling nature of the industry meant that no structure could straddle the entire industry, and therefore, compromises and surrogates were developed. That noted, the CAA penetrated industry much more deeply than the FAA did.

From the time of the GAO report at the beginning of 1998 and EO 13073, Administrator Garvey established a Y2K office and directed it to provide leadership—guidance and oversight—to FAA's seven Lines of Business (LOBs) and aviation industry partners. At about the same time the United Kingdom also started creating several Y2K-related committees that involved aviation. At the departmental level, Department of the Environment, Transport and the Regions (DETR) had a transport committee, which included a director from the CAA and NATS, at which Y2K information was shared. At the CAA itself, the manager responsible for IT chaired a Y2K committee. He reported directly to the CAA Executive Board.

Both the CAA and the FAA relied heavily on external IT service providers for their internal Y2K operations. EDS, the CAA's regular IT service provider, was responsible for ensuring compliance of all level one, two and three systems scheduled for Y2K fixes at the CAA. EDS was also charged with checking with IT vendors to ensure their systems were compliant. There were five audits of the CAA's work on internal systems. The CAA's internal audit department conducted two audits in 1998 and three in total. Deloitte and Touche conducted a fourth audit, with a follow-up, and Martyn Thomas Associates conducted a fifth. In the main, the FAA used contractors to get the work done and auditors to provide assurances. The Y2K Office contracted Science Application International to conduct most of the work. The FAA also contracted a company to provide programme management support.

Like many other large organizations, NATS depended on others to ensure smooth delivery of its service: power, local transport, clean water and so on. NATS's contingency plans did overlap considerably, however, with others' responsibilities. When NATS contacted its suppliers to query their rate of compliance, the response rate was low, and among those who did reply, it was often not verifiable. As a result, NATS had numerous in-house contingency plans, which included work that was already being addressed at the UK/NIF, including preparations for power failure, road transport failure, communications failure. NATS had an internal auditor that reported to the project manager. The auditor checked the inventory, plans and tests (INT 63).

The CAA contacted the 1880 organizations that it regulated through direct correspondence and occasional audits. By way of participating in the UK/NIF, the CAA started its voluntary programme, *Business as Usual*, in which the key players in the industry collaborated to ensure that the industry worked as normal over the millennium period. The 'key players' represented 15 organizations (later 14) (News Release 21 April 1999). As noted, as part of the programme, the CAA contracted AEA Technology Consulting to conduct independent surveys of the participants. Figure 3.4 uses UK aviation as an example to show how potentially densely populated the regulatory space can be in aviation.

In contrast, the FAA did not attempt to penetrate industry as deeply as the CAA did. The FAA held meetings with key players and industry

58

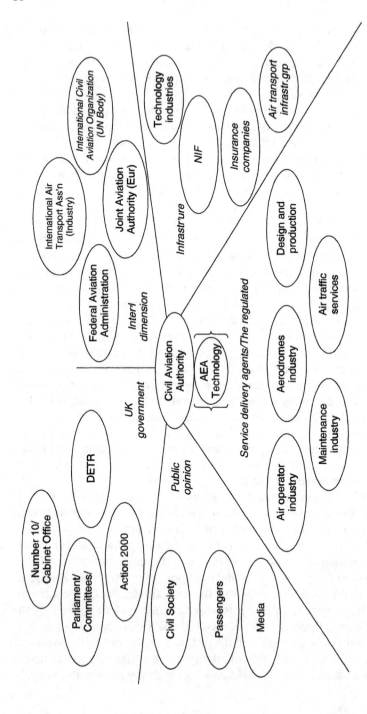

*Figure 3.4* The Y2K-related regulatory space in UK aviation[6][12]

representatives as part of internal initiatives and also as part of the US/WG on Transportation. But the intention of the meetings was not to dictate plans to industry but rather to allow people to share information about what their organizations were doing. The FAA also started opening itself up more—by having more press conferences and inviting its critics to visit the FAA and discuss its plans for compliance. Indeed, while the structure ostensibly was the same in both countries, the style of governance differed, a point explored in the next section.

### The style of the response

*Statistics agencies*

At times moving Y2K compliance projects forward was as difficult as motivating Sisyphus. The work seemed rather thankless: the more work one did, the more work one found one had to do. The problem space was perpetually expanding, and the lack of drama continually made it difficult to inspire the workforce. When emotions ran high it was more often out of aggravation than a sense of triumph.

Formal risk management practices were uncommon at ONS, and it showed. ONS briefing material notes business areas continually squeezed out Y2K compliance work in favour of 'regular' priorities; managers failed to commit to the urgency of the problem; and others did not adhere to deadlines. Many at the ONS had a problem with prioritizing their systems. Formal risk analysis tools and scenario planning were new to most areas. Systems owners had inflated ideas about the criticality of their systems. Indeed, the priority list was rarely stable and only latterly complete—some redefined the criticality of their systems as late as May 1999.

Many operations at ONS were uncomfortable with the high level of scrutiny. In order to get Y2K projects out of their in-boxes many programme areas had rushed, prematurely as it turns out, their systems onto the Y2K-compliance lists. At the same time many were reluctant to disclose the details of their compliance operations. In November 1998, therefore, the ONS executive indicated that each system would have to be compliant by March 1999, and each system would require director-level sign-off. Following this directive, many programme areas were forced to withdraw their systems from the Y2K-compliant list; many systems had been put on the list as compliant when in fact their compliance could not be demonstrated convincingly.

ONS staff had become frustrated by what they felt was excessive external reporting requirements. ONS briefing material noted that during the last 18 months of the programme, the CO redefined and intensified its reporting requirements and that the lines of communication between CCTA, the CO, Treasury (ONS's parent department) and the ONS became very 'confused' at times. They noted that the centralized reporting requirements from the CO did not accommodate 'on the ground' operational nuances. The reporting requirements resulted in 'low priority systems...receiving disproportionate attention, despite having contingency plans in place' (ONS, 2000, 7).

Staff at the BLS were equally frustrated with interactions with staff from outside the agency. The BLS staff noted that the systems were too complex for outside auditors to understand and audit in a matter of weeks (INT 27). The strife between the external auditors and the BLS staff was such that senior staff from the DOL had to come in to mediate disputes.

In another case, one director at the BLS recalled being pressured by the DOL into developing a work schedule that 'showed progress'. That is, that each quarterly report would show marked progress over the previous quarter. The director noted that he advanced the anticipated completion date in order to meet the DOL's (and OMB's) request. When the system was not ready in time, the BLS found itself criticized in the quarterly OMB report. This criticism marked one of the low points in the Y2K project at the BLS (INT 27). Relations between the BLS and the DOL became strained; the head of the Y2K project at the DOL was replaced by a lawyer from within the DOL, with limited budget, IT and project management experience (by his own account) (INT 48).

Internal relations were also strained. Like at the ONS, managers could be 'cagey' when reporting their systems' compliance, for instance. While OMB/DOL reporting requirements were frequently criticized, one BLS IT Director noted that the 'percentage completion' reports had their benefits. Up until the time that the percentages were committed to paper, reports were often given informally, orally and in a positive light. He noted once percentages 'were committed to ink', he was able to monitor progress more closely and, to his surprise, he noticed some percentages did not change over time. This lack of progress forced him to ask more probing questions of the programme areas. In some cases, some areas had fallen behind and had not been sufficiently forthcoming (INT 27).

To a certain extent, the professionalism of the IT staff at the BLS was undermined not only by external auditors and staff at oversight agencies but even executives within their own agency. The staff felt that their advice went unheeded. One interim strategy IT staff adopted to control the scope of the Y2K work was to distinguish between production 'critical' and 'non-critical'. They defined production critical as those systems that were required to produce reports for January 2000. All others were considered non-critical because there would be time to fix them after 1 January 2000, if problems surfaced. (Many BLS reports are issued quarterly or semi-annually.) In September 1998, the DOL/OIG informed the BLS it would be judged by the GAO guidelines, which made no such distinction, and therefore it was back to 'audit, fix and test everything' by the deadline. In another example, one of the interview subjects for the International Price System noted that the system had been upgraded in 1995 and made Y2K compliant at that time. He recalled at an early strategy meeting in late 1997/early 1998 among senior managers that he remarked that his systems would not require any Y2K-related work. He said his director noted, given the environment, 'that answer would probably not fly'. He estimates that a good part of one full year for him and his team was spent exclusively on Y2K, with little ever being found that was not compliant (INT 41).

Ultimately, there was also considerable discrepancy at the BLS as to whether or not the programme was successful. One senior manager commented that the political appointees and most of the executive level within the agency felt Y2K was a resounding success, whereas he and others at the mid-manager level, and particularly IT people, felt that the process was excessive given the limited risk (INT 41). There was also little appreciation for opportunity cost, as far as staff resources were concerned. One executive at the BLS who concurred with this view noted the response was so excessive it should be described as a dereliction of duty (INT 42).

*Aviation agencies*

After a slow start the FAA became much more zealous in ensuring that the agency would become 100 per cent compliant—enforcing strict standards and timelines, driven by the Administrator's Office, in what became at times a confrontational and at times a personally rewarding environment for staff. The CAA, on the other hand, seemed to have a more mundane, systematic approach, though there was the occasional conflict with its IT service provider. Both regulators had a somewhat 'light touch' approach to the industry, encouraging members to share information and work towards compliance, though the CAA did involve itself much more closely with industry than the FAA did.

In the main, Y2K was a slow methodical process for both agencies that was extremely well-documented, including formal risk assessments for critical systems. Some exceptions stand out, however. Air traffic controllers are particular about getting things 'just right' with their systems. This often results in re-scoping project plans. And each change can involve extensive paperwork and several senior-level sign-offs. Hence, despite the CAA's efforts to reduce costs on its Y2K plan by embedding Y2K fixes within other IT projects, there was at least one project whose plans were re-scoped so often that the project would no longer be ready in time for the millennium change-over. As a result, the CAA had to pay to apply the Y2K upgrade to the existing system, despite knowing that the existing system would be retired when the project was complete, shortly after 2000. Interview subjects noted a steep learning curve on both sides as EDS and the CAA learned to work with each other, which often resulted in delays and increased costs in the early days of 1998/99 (INT 65; INT 66).

The intervention of Congress at the FAA marked a significant shift in style at the agency. Administrator Garvey started having weekly meetings specifically on Y2K with the Executive Committee at the agency. Garvey told the Y2K office that they should report to her directly and insisted that they name individuals responsible if progress slipped. One interview subject noted it got some people angry and created some friction but the message was clear: people were going to be held to account (INT 29). The entire approach to the problem changed. One interview subject noted, 'it was removed from the realm of the IT geeks and moved to every corner of top management' (INT 55). It was also characterized as a political problem. The Deputy Secretary at the DOT informed his staff that if there were shortcomings it would make the President look bad at the start of an election year, in which the 'Technology' Vice President would be standing for election (INT 55).

In an effort to stem the flow of criticism, the FAA started engaging much more closely with its critics. To start, on 10 April 1999 the FAA had an end-to-end test to which it invited the GAO, the OIG, the Administrator and the press. The FAA also played a 'war games' scenario, which included numerous 'possible failures', and invited the GAO to attend the exercise. During a day in September 1999, it simulated the change-over from 31 December 1999 to 1 January 2000 in a real-time drill. The FAA invited the press to the exercise. The FAA also invited its most severe critics to its offices, management gurus such as Ed Yardeni and Peter de Jager, to explain what the FAA was doing to mitigate the risk. de Jager eventually agreed to fly on a plane during the change-over, as did Administrator Garvey. The CAA also rethought its communications style somewhat. NATS engaged much more directly with *Computerworld*, a magazine that had been openly critical about NATS progress. When NATS did make progress on its Y2K plan, it reported publicly on how much it was doing—usually the bigger the better—rather than on whether or not the bug was materializing in the manner in which some had predicted. Note, for instance, the words of NATS's press release: 'NATS identified and fixed *when necessary* some 700 operational air traffic control and 170 non-operational systems' (NATS, 1999a; my emphasis).

With respect to the aviation industry, the CAA first learned of Y2K from one of the major airlines. Prior to Number 10's intervention, the CAA approached Y2K as it did any other safety issue, though it recognized Y2K was more obscure than problems with which it was accustomed to dealing. The CAA aimed to take reasonable steps to ensure that the mitigation of Y2K-related risk was carried out in a coordinated fashion and to acceptably safe standards. In a memo to the industry, the CAA indicated that the prime responsibility for ensuring continued safe operation lay with operators and suppliers of civil aviation services and products:

> In short, the Year 2000 problem is a safety issue like any other... in [its role as regulator], CAA [Safety Regulation Group] is requiring all operators provide safety assurances of Year 2000 readiness. Areas should include as a minimum—issue recognition; impact analysis; Year 2000 planning; contingency planning; accomplishment statement. (CAA, 1998)

Ultimately, as part of its *Safety Programme*, the CAA concluded that 1858 regulated organizations demonstrated they were Y2K compliant; 22 did not. The 22 that were not compliant the CAA did not consider to be a risk to public safety and, therefore, elected not to take any precautions with respect to those companies.[13]

Similarly, despite the government publicly placing such importance on Y2K compliance and the desired target of 'no material disruption to service', *Business as Usual* was voluntary, and results between the auditor and the audited had to be agreed before they were finalized. Indeed, one organization was able to withdraw from the programme. The organization in question was undergoing a reorganization of its company's reporting structure and it felt it could no longer meet the CAA's timelines. The company withdrew from the UK/NIF with little fanfare. As far as the UK/NIF was concerned, the CAA simply counted 14 instead of 15 companies.

After pressure from Congress, the FAA also started working much more closely with industry. One interview subject noted that industry had been working previously in silos (INT 38). Most organizations had highly developed Y2K plans but were not sharing them with others. By the end of 1998 the FAA was sponsoring industry days and having leading players in the industry present their Y2K plans to others. The interview subject noted it was cooperative.

## Conclusion

Both governments' responses to Y2K within departments and agencies encapsulate the chief characteristics of Bureaucracy: hierarchy, specialization and standardization (Aucoin, 1997). Direction was set from the top; IS specialists within programme areas were responsible for fixing their own systems; and all units followed (by and large) the same process (inventory, fix, text and audit).

There were considerable strengths to such an approach. Uncertainty loomed large, particularly in the early stages. As we shall see, 1 January 2000 was described in an almost Armageddon-like fashion. Agencies were slow to report and reported inconsistently. Government, and indeed society as a whole had come to depend—or so it seemed—on multiple, interdependent and complex technologies. The selected approach brought discipline to the exercise. It defined (broadly) a process; it set timelines and articulated outputs. This standardization helped ensure that the Executives' promise would be met: critical services would not suffer Y2K-related failures.

Nevertheless, the episode elicits concerns about the functioning of bureaucracy under these conditions. Indeed, there is an irrational side to this rationalist approach. First, the age-old complaint about bureaucracy: inertia. It is remarkable to consider that organizations such as the case studies described here—all almost entirely dependent on technology to deliver their respective services—failed to act corporately on Y2K much before 1998. The FAA received by far the most criticism and public scrutiny but in fact they were all guilty to some degree of neglecting their critical infrastructure and severely compromising their capacity to deliver their services that they are mandated to deliver.

Second, the failure to align service accountability with systems accountability made it difficult to hold people to account for the 'Y2K compliance' of their services. In some respects the Executive orders assumed the characteristics of the public sector reforms of the day: people would be held to account for service delivery, which, it was presumed is what citizens care about most, according to the Reinventing Government and New Public Management reforms of the time. However, these organizations were a myriad of hardware and software technologies whose interdependencies—and accountability—could not easily be mapped. Those who were responsible for particular services were not necessarily responsible for the technologies on which their services relied and therefore were unable to speak with confidence about the compliance of the service. Bureaucracies do not lend themselves to lateral exercises, of course. Accountability goes up, not across. Initially, at least, there were few fora or incentives to pursue such cross-organizational work. Hence, the exhaustive, system-by-system approach was the only way to ensure that all systems and therefore all services would go undisrupted.

Third, middle management was severely under-utilized. The hierarchical ladder represented authority not expertise. So while the higher up the ladder one went the more responsible one was for the broader mission of the department, the less aware one was of the individual component systems that supported the work. Again, it was safer to simply order that everything be fixed. A critical rung in this ladder, however, is the role of middle management—a key link between the front-line specialists and those who are ultimately accountable for the successful delivery of the services. Yet middle management was used not in an advisory role but rather as a postbox: they frequently assembled reports and sent them up the line. They had plenty of work, and pushed the process forward. But their capacity to fuse system-specific knowledge with the broader mission of the organization was an intellectual resource that went frequently untapped. This, in many respects, made the project less more work than perhaps was necessary.

Fourth, while many argued that leadership was crucial to overcoming the inertia of the organization, questions of leadership remain unresolved. The leadership style and even some of the structures the leaders put in place were different yet these differences seemingly had little impact on the outcome. While testimonies about hard work and commitment by FAA staff spurred on by the Administrator are compelling, and contrast significantly with the scepticism one encounters at the BLS, for instance, both agencies finished, largely on time and by the same method. Indeed, the strict and prescriptive CO/OMB guidelines left little room for interpretation at the agency level.

The best we can say is that Congress made a difference and so did the White House, Number 10 and Cabinet Office. The Y2K episode suggests Congress can have a powerful impact on management initiatives in government. GAO reports and Congressional committee hearings and report cards brought considerable attention to the subject—from the media, the public and the Executive Office, which in turn had a significant impact on how government departments and agencies responded to the problem. In many respects the turn-around at the FAA is a phenomenal public management success story brought about by focused leadership at the political level.

On the one hand, this tells us that strong direction from the top is potentially crucial for public management success in a problem of this magnitude. On the other hand, the tools overseers employ in their interventions can have perverse effects and should therefore be used only after careful consideration. Letter grades in high level report cards and standardized templates often neglect the subtleties embedded in complex problems. Earning approval from Congress and Cabinet Office frequently came at the cost of very expensive Y2K plans that aimed to eliminate many aspects of the risk rather than manage them.

While this approach may have worked to some degree within government departments and agencies it was much more difficult to achieve across the critical infrastructure: not only was the system too complex but authority was diffuse. Still, there was a similarly rationalist impulse, particularly in the United Kingdom. The UK/NIF is an effort to map (and to a certain extent control) the entire infrastructure. While ambitious, such a move was not entirely possible: there were clearly several trade-offs embedded in the forum, especially with respect to national and international supply chains, the voluntary nature of the forum and somewhat opaque nature of the reporting, particularly in the United Kingdom whose approach was more cohesive but also perhaps more compromised.

This chapter has focused on a brief period of time and a limited number of agencies. To focus specifically on these cases in isolation is too narrow. To a large extent the seeds of what became known as Y2K were planted years earlier, and would grow to make almost every organization vulnerable to collapse, or so it seemed. To understand the context in which these organizations were working in the late 1990s we turn first to consider the exponential growth of IT, and in particular decentralized IT, right across Western markets, the launching point of the Y2K story and our next chapter.

# 4
# The Market Failure Hypothesis

Early forecasts concerning the impact of Y2K were bleak; however, the true magnitude of the problem will never really be known. From a technical standpoint, Y2K was largely characterized as a problem of dependence on IT, including embedded systems; interdependence across systems; systems complexity; and a fixed time horizon. These assumptions are contestable but not refutable.

The uncertainty and perceived uniqueness surrounding the problem precluded any actuarial modelling, which made it difficult to insure against. Legal liability was similarly unclear though ultimately few Y2K lawsuits were ever filed. Given the time constraints and the magnitude of the problem, it made more sense to cooperate with IT companies to find solutions. In fact, the US government in particular legislated to ensure it was even more difficult than normal to pursue organizations in court over IT matters.

The cost of Y2K was exaggerated; many tasks and efficiencies, for instance, were achieved under the aegis of Y2K but in fact had little to do with the bug. That noted, a precautionary approach within government departments and agencies made most Y2K projects resource intensive and went beyond any measured approach. It was a boondoggle of the first order.

Across the infrastructure as a whole, both governments played an information-sharing and awareness-raising role that filled a gap that emerged at the interface of organizations, particularly among those in different sectors. Given the high degree of awareness of the problem, the United Kingdom's high-level of standardization across the infrastructure, while ambitious, makes less sense from a market failure perspective than the American intervention.

This chapter starts with a brief discussion of what Y2K specialists were predicting concerning Y2K. It then drills down: it examines the technical nature of the risk, the role of insurance and the law in the face of Y2K and the market as a regulating force. It concludes by examining whether both or either governments' responses can be explained as a response to a market failure. First, we shall review the Y2K *seers'* predictions.

## Y2K forecasts and failures

Gartner Group was the leading commentator on Y2K. Gartner is an IT research company known for selling assessments and visions of 'where technology is going' (Feder, 1999a). Gartner assumed a leading role on the issue when in 1996 it testified to the Subcommittee on Technology and the Subcommittee on Government Management, Information and Technology that Y2K would cost the US economy $300–600 billion (Feder, 1999a). The *Sun Journal*, a publication from a major American IT manufacturer, attributes the following predictions to Gartner Group analysts:

— Seventy per cent of applications would experience some type of failure;
— Thirty per cent of mission critical systems would experience some type of failure;
— Ten per cent of businesses were likely to fail;
— It would take one person-year to fix every 100K lines of code ('Year 2000 Presents IT Organizations with Challenges and Opportunities', *Sun Journal*).

Companies like Gartner had used questionable methods, however. Gartner's 1996 costing forecast for the US economy was based on putting Gartner estimates of the lines of computer code in use together with the average cost of hiring outside programmers to fix faulty mainframes. It was later acknowledged that this was too crude a measure (Feder, 1999a). Moreover, when Gartner made forecasts about systems failures, it relied heavily on unverifiable reports by computer users of their plans and progress (Feder, 1999a).

Gartner was not alone in its gloomy outlook, however. Ed Yardeni, Chief Economist at Deutsche, Morgan Grenfell, famously predicted that there was a 70 per cent chance of a 12-month recession starting

1 January 2000 (1998). Peter de Jager, a Canadian consultant who made his name on Y2K by penning the first 'popular' article on the subject, 'Doomsday 2000' in *Computerworld* in September 1993, predicted 1 per cent of companies would go bankrupt as a result of the millennium bug (Taylor, 1997).

As for actual failures, many appeared well before 1 January 2000. Capgemini, in its ongoing survey of information technology executives at 161 companies and government agencies, reported in August 1999 that 75 per cent had reported some Y2K-related failure (Hoffman, 1999). However, only 2 per cent of the companies polled suffered business disruption because of the problem; most worked around the glitches. In the US Senate report *Y2K Aftermath—Crisis Averted: Final Committee Report* (2000) the committee appended 13 pages (283 examples) of Y2K-related glitches that actually occurred, organized by sector: utilities; healthcare; telecommunications; transportation; financial; general business; general government; international. The UK government declared in a news release that there were 82 minor glitches across the UK government (Grande, 2000).

The Senate report underscored that the extent of the problem would never be known because only a small fraction of the occurrences would ever be reported. There was no incentive for countries or organizations to report computer problems. As with any internal problems, organizations were likely to fix them and continue their operations unbeknownst to the general public (United States Senate Special Committee on the Year 2000 Technology Problem, 2000, 37).

It is also worth noting, however, that intervention also caused problems. When programmers open up code to fix it, given the tenuous nature of code (for example, designed over several years and by several programmers and not very well documented) programmers run the risk of entering new bugs into the system. And indeed, there were cases in the Y2K story of upgrades and testing procedures causing greater problems than the ones they tried to fix. (See, for example, Delaney, 1999; Feder, 1999b). Problems of this nature were rarely noted by interview subjects when discussing Y2K strategies.

## The technical nature of the risk

### Framing the problem
The magnitude of the technological problem was largely framed by four (overlapping) assumptions and/or descriptions, which I will develop

in this section: (1) growing dependence on IT; (2) the complexity of systems; (3) the pervasive nature of embedded systems; and (4) organizational interdependence.

### Growing organizational and social dependence on IT

Both countries had a growing dependence on IT; however, the dependence in the United States was always far greater. Not withstanding the fact that there were incentives to exaggerate Y2K expense claims in the United States, in particular, as noted in the previous chapter, the degree of penetration of IT into the US economy made the US government more vulnerable to expensive Y2K operations. In 1983, Collier's Encyclopaedia estimated there were two million PCs in the United States. By 2000, the Farmer's Almanac estimated there were 168.9 million systems—a growth rate of 30 per cent per annum. Indeed, the IT sector in the United States was estimated at $4.6 trillion by the year 2000. At the time in question, the rate of growth in IT in the United Kingdom was significant, but consistently lower than in the United States. Table 4.1 and Figure 4.1 below summarize the expenditure and increased dependence on IT at the time.[1]

### The complexity of systems

Despite the seemingly simple and cursory manner in which the Y2K problem was frequently framed in the media, the problem could often be deceptively complex. This complexity was the result of the intricacies in individual systems and the proliferation of systems outlined above. Even within the individual systems, date-related problems could arise

*Table 4.1*  IT market and GDP (Excerpt)

| Country | IT spending as a percentage of GDP | IT spending as a percentage of worldwide IT spending | | IT Market compounded annual growth rate |
|---|---|---|---|---|
| | 1994 | 1987 | 1994 | 1987–1994 |
| US | 2.8 | 44.8 | 41.4 | 8.7 |
| Europe (including the UK) | 1.6 | 29.1 | 27.6 | 9.1 |
| UK | 2.1 | 5.5 | 4.7 | 7.6 |
| Others | | 26.1 | 31.0 | 12.6 |
| World | 1.8 | 100 | 100 | 9.9 |

*Source*: OECD, IDC and the World Bank, as cited in Cartiglia *et al.*, 1998.

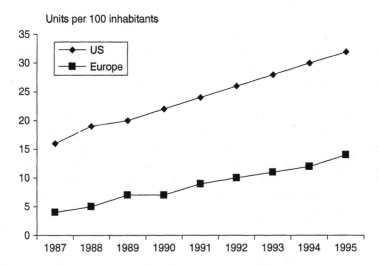

Units per 100 inhabitants

*Figure 4.1*  PCs installed in the United States and Europe
*Source*: OECD, based on ITU (1995) data, as cited in Cartiglia *et al.*, 1998.

from within systems from hardware, software and/or data. The necessary interactions between these three made it difficult to predict which rule from which system would predominate. Hardware and software were frequently developed by different manufacturers and data could be input and manipulated in a variety of ways. With no agreed standards on 'date entry', it was impossible to say for certain how date fields would be interpreted. Similarly, with bigger systems, such as those at the FAA, it was common to have a programme that contained *millions* of lines of code. Understanding exactly how this code ran was a monumental task; it was certainly beyond the scope of one programmer.

Some argued the problem existed only for old mainframes and legacy systems but this was not strictly so. The proliferation of systems, particularly PCs, also made the problem more complex. With the advent of user-friendly software (for example, Access, Lotus, Excel) and the relative ease with which one could obtain it (for example, local shop, download from the net, copy it from one's neighbour) people started developing their own programmes within these software applications. Practices that seem common today were less recognized at the time. For instance, accountants and economists developed spreadsheets to track budgets; statisticians developed forecasts; managers tracked staff hours and project progress; staff managed their own word processing. And, of course, files were shared readily between individuals by disk, CD, e-mail or shared server. Executives actively encouraged this behaviour by making the PCs, Internet access, software and relevant training available to their staff. But this dynamic environment occurred largely outside of the control of IT departments within organizations. In the five-step process in managing a Y2K project, 'inventory' was always the first step and often the most time consuming because IT departments rarely knew all the systems in the organization; nor did they necessarily know how they all worked, whether they had any date-functionality or how critical they were for programme delivery. With each of the agency case studies, IT staff noted they discovered during their Y2K inventory systems that they did not know existed but were essential to their departments for service delivery.

### The pervasive nature of embedded systems

Embedded systems shifted the problem even further away from IT departments. Embedded systems contain programmed instructions running *via* processor chips. The processor chips are similar to stand-alone computers buried inside various kinds of equipment and are particularly common in manufacturing systems. In 1998 OECD estimated there were 25 billion embedded systems worldwide (OECD, 3). Action 2000 estimated that 7 billion were distributed in 1996 alone (House of Commons Library, 1998, 10). The number of embedded systems, coupled with their prevalence in equipment, fuelled speculation about massive systems failures. Ann Couffou, Managing Director at Giga Year 2000 Relevance Service, in 1997 testified to Congress that the following areas and systems were vulnerable because of embedded systems: manufacturing control systems; elevators; telephone systems; medical equipment; stock markets; military messaging systems; radioactive material waste systems; fax machines; electronic time clocks; landscaping systems; vending machines; thermostats; and microwave ovens. She concluded: 'anything with an electronic component should be suspect. The rule should be guilty until proven innocent' (Couffou, 1997). Indeed, the embedded systems problem is foregrounded in most government publications.[2] When the issue of embedded systems came to the fore, the problem became even more complex organizationally because it involved new players in new ways. Y2K projects started to include building managers and property management companies (that is, 'Will our elevators work?' 'Will our doors open?') and the manufacturing industry ('Will our assembly lines work?' 'Will our products that roll off the line work?') The solution to embedded systems was similarly complex. Many people did not know if their systems contained embedded systems and they almost certainly would not have known if the programming involved contained date-functionality. One would have to contact the supplier—if indeed the supplier even knew and still existed.

*Organizational interdependence*

The GAO (1999a, 10) described the tenuous interdependence dynamic and the Y2K threat thus:

> Our nation's reliance on the complex of array of public and private enterprises having scores of systems interdependencies at all levels accentuates the potential repercussions a single failure could cause.

The interdependence argument understood broadly could be cut several ways. First, there is electronic interdependence, which is to say simply that electronic systems can depend on each other due to their interconnectedness. Second, at the organizational level there is also interdependence between suppliers. The major American automotive manufacturers, for instance, estimated that they had collectively 150,000 suppliers on whom they depended (Sendle, 1998). Third, interdependence can exist across governments, joining the sub-state with the state, and the state with the supra-state organization, as well as all the combinations and permutations that exist therein.

To turn to the agency case studies, word length limitations prevent a full discussion of the complexity of Y2K in technological terms in the industries. The following examples are meant to be illustrative. Air traffic control is an extremely complex system which dwarfs almost all other Y2K operations. NATS bespoke systems are extremely large and complex. Its Y2K operation involved examining 700 operational air traffic control and 170 non-operational systems (NATS, 1999a,b). NATS' three air traffic control centres manage 5000 flights passing through UK airspace, as well as over the North Atlantic, every 24 hours. Moreover, the international component was also critical, for the United Kingdom in particular, where 75 per cent of all flights were international and therefore relied on the safety practices and regulations of other countries.

The complexity, however, does not derive solely from the volume of the technologies but also from the inconsistent manner in which they are maintained. Despite its dependence on IT, the agencies' track records for document management are sketchy. At the start of the Y2K project, neither the CAA nor the FAA had a reliable inventory of systems. Eventually the CAA identified 120 'systems' and in addition noted 40 + standard and specialist software products as well as an inventory of over 100 types of equipment. There were many Access databases and Excel spreadsheets, for instance, that had been developed over time, but the IT department had not known that they existed.

NATS in particular had a poor track record for documenting system changes. This is partly because paperwork is viewed as 'boring' and partly because there is so much paperwork to do. NATS frequently carries out detailed probability risk assessments, for instance, on changes to systems. One interview subject noted that NATS' logs can frequently be eight or nine months out of date (INT 63). This was particularly worrisome at the aviation agencies because their IT is *not* COTS (Commercial Off-the-Shelf) and therefore could not simply be replaced by a new version. Almost all technology is reworked by agency staff once it is purchased and therefore fixing it sometimes means opening up the source code. This can be a lengthy process.

But this does not mean that the aviation agencies were necessarily discovering many Y2K-related problems. Air Traffic Control (ATC) was perhaps the area that received the most attention. Ironically, ATC staff spent most of their time on the administrative systems. One interview subject noted that the HOST system, which runs the ATC, ran on a 32-year calendar and was not going to have a problem until 2006. In addition, although many systems with date functionality helped the operation run more smoothly and could have had an indirect health/safety impact, they were not absolutely critical. For the most part, ATC staff could have reverted to manual processes, if necessary (INT 44).

The private aviation industry also had a complex environment, which was highly dependent on technology and extremely interdependent. A typical 747, for instance, can contain as many as 16,000 embedded chips. Moreover, on the ground, there was a proliferation of systems upon which the industry depended: finances, bookings, payments, staffing schedules, baggage handling, pilot licensing, seat assignment, security and access, parking, communications, monitoring of runways and train links (GAO, 1998a,b). The American airline, Delta, for instance, said that of its 600 systems at its headquarters in Atlanta, almost half were considered mission critical (Hodson, 1998, 6).

The statistics agencies were less dependent on external service providers but were completely dependent on technology to provide their service. As one interview subject from the BLS noted, every step that the BLS takes in its mission is done electronically—data collection, manipulation and reporting (INT 27). The ONS is virtually the same. ONS inventory indicated in the run up to 1 January 2000 that it had 107 'High Priority' systems, 146 'Medium Priority' systems and 152 'Low Priority' systems.

Responsibility for these systems at the ONS was dispersed, making it more difficult to coordinate and obtain reliable information about each system. Responsibility for the 107 'High Priority' systems, for example, was distributed among 82 different systems owners. As noted, briefing material notes that it was difficult to establish a priority list because staff kept changing the ranking of priority systems, according to which ones were priorities for a given month. Relatedly, Y2K staff noted that it was difficult to get time on high priority systems because they were frequently in use for their operational purposes. (Y2K work usually required taking a system off-line.) The ONS argued that progress on Y2K slipped due to lack of resources, the expanding nature of the task and the unrealistic timelines.

BLS staff attributed the lack of progress to the complexity of the task; the extensive paperwork and audit trail that was required; the 'unrealistic' reporting requirements; the slow, end-to-end testing that had to occur before systems could be guaranteed; and the extensive testing that occurred on systems that were not vulnerable but had to be tested for audit purposes (INT 27–8; 35–6; 37; 41–2). BLS staff argued that processes were frequently made up of several 'subsystems'. A *system* would not achieve full compliance until all *subsystems* had achieved compliance. This made the Y2K compliance project slow,—not just because all of the subsystems had to be tested but because the interfaces between these subsystems had to be tested too. As a result, when one reviews OMB Y2K reports on the BLS one does not see a slowly increasing percentage of systems being fixed but rather significant step changes across the organization around the same time. This explanation goes some way to explaining why the Department of Labor made seemingly negligible progress on mission critical systems between May 1998 and November 1998 but made a significant leap between November 1998 (67 per cent) and March 1999 (85 per cent). This approach, however, frustrated staff at the OMB who wished to see and show progress over time: it made it more convincing to Congress and the public from a reporting standpoint.

### Technical solutions

Defining just what exactly was meant by 'Y2K compliance' was difficult. What started as an *ad hoc* approach to fixing a local problem soon became a more complex struggle given the various interdependencies that existed between organizations and systems. In response to demand from the public sector, UK industry, and commerce in particular, the British Standards Institution committee, BDD/1/-/3, developed a four-rule definition of Y2K compliance. BT, Capgemini, CCTA, Coopers & Lybrand, Halberstam Elias, ICL, National Health Service, National and Westminster Bank all contributed to the definition. The definition is as follows:

> Year 2000 conformity shall mean that neither performance nor functionality is affected by dates prior to, during and after the year 2000. In particular:
>
> Rule 1. No value for current date will cause any interruption in operation.
>
> Rule 2. Date-based functionality must behave consistently for dates prior to, during and after year 2000.
>
> Rule 3. In all interfaces and data storage, the century in any date must be specified either explicitly or by unambiguous algorithms or inferencing rules.
>
> Rule 4. Year 2000 must be recognized as a leap year.

From a technical standpoint there were many possible approaches to fixing the problem. Five were particularly common: date expansion (increasing the year field to four digits); date compression (storing dates in binary or packed decimal formats); windowing (splitting the century into past and future windows, either fixed or sliding); encoding (recoding years to a new numbering scheme); and encapsulation (shifting the date backwards on internal clocks, usually by multiples of 28 years) (IBM, Prickle and Associates, OECD, 1998, 10). Some approaches were more demanding and thorough than others (date expansion was considered not only the most time-consuming but also the most reliable) but all were considered labour intensive. Although many members of the IT community acknowledged that older mainframes were the most vulnerable to Y2K-related failures, Y2K-related glitches existed in hardware, software and data until the mid-1990s and thereafter. Windows 95, for example, was one higher profile and commonly used operating system that was not considered Y2K compliant (King, 1999). It should also be noted that there was no 'one size fits all' option, what was referred to as 'the silver bullet' solution. It was possible to install software on a system that would check for date-related problems on standard systems—standard desktops, for instance. But in a highly decentralized environment in which people created their own programmes this alternative was not considered foolproof.

## All doom and gloom? Some alternative views

Many countries did considerably less than the UK and US governments yet experienced no significant systems failures.[3] Many local governments[4] and small and medium-sized enterprises (SMEs)[5] were also considered far behind in Y2K compliance, and they too experienced few failures. Manion and Evans (2000) argued that most countries were less dependent on technology than the United Kingdom and the United States. They also argued that many countries had made considerable progress in the last six months of 1999. Both points are no doubt true. *Most* countries are not as dependent on IT and *some* countries did make advances in those last six months. Nevertheless, if the bug had the bite that government publications—and particularly the earlier publications in 1997/1998 proclaimed—one would still expect to see a disproportionately high number of failures in countries that prepared less. But instead, there are very few failures and almost no significant ones. The following discussions—dependence, date functionality (PC testing example), embedded systems and systems interdependence—illustrate some examples of questionable assumptions that informed seemingly exaggerated responses.

*Dependence on IT*

Despite one consultant commenting, 'the more you looked the more you found', one might just as easily suggest the opposite, that is, the more one looked the less one found. It depended on where one was looking and what one's expectations were. For instance, despite being the largest business group with 84 per cent of employee organizations, micro-businesses in the United Kingdom were found to be low risk. The Standing Committee on Science and Technology in 1997/1998 estimated that 10 per cent did not even own a computer. By late 1999, Action 2000 revised that number upwards to one-third of all micro-businesses. Of those that did own a computer, most were eventually understood to be isolated, not interdependent. The state of play among SMEs varied depending on the sector. In the finance sector, for example, all were considered exposed to Y2K-related risks, but the perceived risks declined in other sectors. In retail 7 per cent of organizations did not operate any IT or process plant equipment; in agriculture, it was 8 per cent. In transport and logistics, 19 per cent did not have IT or process equipment, and in travel and restaurants, 27 per cent did not. However, this data was not available until early in 1999 (Action, 2000, 1999).

American small- and medium-sized businesses were more dependent on technology but there seem to have been few Y2K-related failures there also. The February 1999 Senate report quotes the National Federation of Independent Business: '330,000 firms risk closing their doors until the problem is fixed and more than 370,000 others could be temporarily crippled' (1999, 128). Similarly, though less dramatically, Koskinen notes in his final report that survey evidence suggests that 70 per cent of companies reported some Y2K-related problems before the roll-over but the public would never learn of them because they were fixed in-house with little fanfare. Ultimately, he concludes that lack of information about what small businesses were doing was an ongoing challenge because of the magnitude—there are 23 million small businesses in the United States (President's Council on Year 2000 Conversion, 2000, 18).

Neither the US Government nor the UK Government seems to have acknowledged organizations' and individual's ability to adapt to changing circumstances or to respond in the face of a problem. It seems that many programmers did so in the run up to the date change. For example, despite its best efforts, the ONS did have one Y2K-related failure, *viz.*—death certificates. Yet the failure of the system did not stop people from dying nor did it stop the ONS from registering that fact. The ONS staff fell back to a method, tried and true—*la plume*. One Y2K consultant noted that many people—especially in the manufacturing sector—simply restarted their systems after an initial failure at the date change without a problem. Interestingly, every US and UK governments' Y2K *post-mortem* observes in the lessons learned section that Y2K brought to light the significant extent to which their respective economies are dependent upon IT. *Post-mortems* rarely, if ever, note the variety of dependence across sectors or the ability of organizations and people to respond with flexibility and creativity to system failures.

### Date functionality

One organization in this study had a Y2K plan that required that each desktop computer be tested for Y2K compliance with a specialized software package. This was common practice. The testing process was labour intensive (that is, each computer had to be tested individually and each test took approximately 20 minutes to run). Of the 98 computers tested, 42 per cent failed. However, if one considers only the computers purchased in 1997, 1998 and 1999, the failure rate was just 4 per cent.

This is not surprising. Despite the claims from Y2K consulting services, such as Greenwich Mean Time (GMT), that the bug could strike from anywhere (and its testing suggested failure rates of upwards of 50 per cent[6]), pre-1997 systems were much more likely to fail Y2K compliance tests. IBM, for example, testified to the Standing Committee on Science and Technology in 1997 that its computers were Y2K compliant by 1997 (Chapter 4, 2). Microsoft, for its part, issued guidelines about which of its software packages exhibited problems with date change. Again, software issued in 1997 or thereafter was judged to be at extremely low risk (Haught, 1998, 5–7).

However, Y2K consultants frequently took a very narrow view of what 'Y2K compliance' meant. They did not accept, for example, the 'windowing' technique (noted in the Solutions Section), which allows programmers to apply *one* rule to programmes to cut back on the time required to fix every line of code. In brief, 'windowing' meant that programmers could pick a cut-off year—say 2029—and programme an instruction that interpreted any year entry that was '29 or lower' as the twenty-first century and any year entry that was '30 or higher' as the twentieth century. GMT described 'windowing' as allowing the computer to 'guess' which century users meant when they entered data. In fact, windowing is a common practice in programming in general and was often used with success when dealing with Y2K (INT 57).

### Embedded systems

How likely were the embedded systems to fail? The government's advice was imprecise. Citing research from 1996, the UK government reported that 5 per cent of embedded systems were found to fail millennium compliance tests. More sophisticated embedded systems were reported to have had failure rates of between 50–80 per cent. In manufacturing environments, Action 2000 reported an overall failure of around 15 per cent (House of Commons Library, 1998, 10).

Without a more specific definition of what 'failure' means in the Action 2000 documents and just how the Y2K probabilistic risk assessment tests (PRA) were carried out and on which systems, it is difficult to conclude meaningfully on the likelihood that embedded systems would fail. The Government's own PRA calculations were never made public. This failure on the part of government is surprising given Prime Minister Blair's commitment to keep the project as transparent as possible (Blair, 1998).

Judging by the outcome, the embedded systems problem was exaggerated. In late 1998 and early 1999 the US Government's Senate

Committee on Y2K estimated an embedded chip failure rate of between 2 and 3 per cent. By late 1999, the committee revised its estimation to a failure rate of 0.001 per cent—a reduction by a factor of 2000–3000 (United States Senate Special Committee on the Year 2000 Technology Problem, 2000, 10). In a discussion with one professional staff member from the committee who helped draft the report, he indicated that the revised figure of 0.001 per cent was an estimate-guess. It meant simply to indicate that the committee's fears of embedded systems had been grossly exaggerated in the early days and that as the date-change approached its fears had diminished (INT 39).

Finkelstein points out that 'only a tiny portion of [embedded] systems have a clock at all and a very small subset of these use absolute as distinct from relative time' (2000, 2). One extreme example comes from the aviation industry. In its testing of 747s, Airbus reported that only four embedded systems were found with any date/time functionality out of a possible 16,000 embedded systems found on an aircraft. None of the four embedded systems functioned in critical systems. In fact all were in the in-flight entertainment system. (Hodson, 1998, 6). So while one's movie may have crashed one's plane was certainly not going to.

In its testimony on Y2K to the Standing Committee on Science and Technology, the Health and Safety Executive (HSE) warned against overreaction with respect to embedded systems. The HSE noted that one health trust had tested over 7000 pieces of medical equipment and 200 had some type of date/time functionality, or about 3 per cent (1997/1998, 4, 4). (The HSE did not indicate whether or not the date functionality in these systems was critical to the operation of the system or what kinds of medical equipment they examined.) The Food and Drug Administration echoed the HSE's position on medical devices, though not until much later, concluding that problems had been overstated (Middleton, 1999). Similarly, despite apocalyptic claims of pending nuclear disasters on 1 January 2000, HSE reported that high-risk nuclear installations' safety-critical systems were not time/date dependent (Chapters 3, 5). But for all intents and purposes, these warning were ignored; they did not figure into the committee's conclusions and recommendations. Nor are these points picked up in any of the NAO reports. The Science and Technology Committee report argued that statistical evidence of failure was of limited use because—as the GAO quotation from earlier in this chapter indicates—even the smallest failure could cause problems; many government systems were so interdependent that one failure could have a ripple effect, a point considered in the next section (Standing Committee on Science and Technology, 1997/1998, 2, 2).

*Interdependence in government systems*

There is a considerable degree of interdependence between and among systems, organizationally and electronically. The UK/NIF and the Working Groups capture a bird's eye view of the network. However, the committee's fear of systems' interdependence belies the actual state of interconnectedness in government systems and fails to distinguish between tightly and loosely coupled systems (Perrow, 1999, 4–6). Some of the limitations to the interdependence argument bear noting. Margetts (1999, 47) in her study of information technology in the US and UK governments between 1975 and 1995, and Fountain (2001, 29–30) in her study of technology in the US Government note that the vertical integration of the governments undermine efforts at horizontal integration. Labour's *Modernising Government* (Cm 4310, 1999) and Gore's *From Red to Results* (1993) acknowledge this limitation, for example, in their respective calls for more 'joined up' (UK) or 'integrated' (US) service delivery. In fact, in the United Kingdom, policy initiatives over the past 20 years, such as privatization, Next Steps agencies (Efficiency Unit, 1988) and Market Testing (Cm 1730, 1991), created barriers that made it more difficult to join-up services across departments, or even units.

The difficulty in joining up services *electronically* had been further exacerbated by the weak IT bodies at the centre, such as the poorly funded and staffed CCTA in the United Kingdom (Margetts, 45). Margetts observes that CCTA's funding was so often the target of cutbacks in the 1980s and 1990s that the office became more reactive than proactive until such time as strategic IT decisions were eventually shifted from CCTA to the Treasury. The state of central coordination in the United States was no greater than that found in the United Kingdom. Just as the United Kingdom began coordinating IT activity at the Cabinet Office the US government started coordinating common IT activity for the first time at OMB (Fountain, 2001, 30). In reply to 'Reinventing Government' initiatives and *The Clinger Cohen Act*, US government departments and agencies initiated a more 'joined up' approach to IT with the creation of an executive CIO position in every department and the government-wide CIO Council. In fact, Y2K was emerging at the same time as the Clinger-Cohen Act. Y2K could be seen as the first big 'test case' for the CIO Council (President's Council on Year 2000 Conversion, 2000, 4). Little by way of coordination had occurred up to that point (Fountain, 2001, 197). Departments were more independent, at least from other government bodies, than the Y2K rhetoric of the day would suggest.

At the individual systems level, even when systems are joined up, it does not necessarily follow that one system shutting down will lead to another one shutting down. One advantage of wide area networks (WAN) and distributed processing is their capacity to resist the knock-on consequences of systems failures. By providing integrity checking at interfaces and independent computational resources, such systems are inherently much less failure prone than other conventional systems. Ripple failure is a very rare failure mode (Finkelstein, 2000, 3). Again, the kind of evidence that characterized the probability of failure as either negligible or extremely low probability did not find its way into government publications or planning.

With respect to the agency case studies, again one sees as much emphasis on independence as interdependence, depending on where one looks. The Department of Transportation staff acknowledged the degree of agency independence the FAA enjoyed (INT 55). The FAA is in a rather unique situation; it develops large systems, over several years, during which time the technology invariably changes, which requires an update to plans and a subsequent re-profiling of the finances. Because of this situation, the FAA, backed by statutory exemption, is afforded considerable independence from the Department of Transportation and Congress with respect to its IT planning. Similarly, the statistics agencies pride themselves on being independent (INT 48; INT 21). Their reports must be seen as being free from any government interference and therefore they too have flexibility in developing plans and strategies for their own technology.

## Insurance, the law and Y2K lawsuits

In 1997 and early 1998, insurance industry commentators speculated that Y2K-related lawsuits could amount to anywhere from $400 billion to $1 trillion dollars in the United States.[7] The first Y2K-related lawsuit was filed in mid-1997 by an SME and settled out of court for $250,000 in compensation for damages (Adams, 1997a).

Around this time the insurance industry took steps to limit its exposure to Y2K-related lawsuits. In the United States, large insurers of manufactured and electrical goods, such as Cornhill, drafted exclusion clauses into policies to protect themselves from liability (Adams, 1997b). Similarly, 25 state insurance regulators approved wording for general liability policies excluding any claim for losses related to Y2K (Adams, 1997d). Partly as a result of these efforts and the *Y2K Act*, which would follow, by spring 1999 analysts reduced their Y2K-related lawsuit forecast and claimed that the US insurance industry would likely pay out $15–35 billion related to Y2K (Lohse, 1999a). In the United

Kingdom, the Association of British Insurers (ABI) was prompted by an independent report that noted that 75 per cent of British companies had coverage on risks to business disruption (Adams, 1997c).[8] The ABI argued that Y2K was a foreseeable and preventable problem and was therefore not an insurable risk. ABI warned companies publicly that they could not expect to be covered for Y2K-related problems and that the onus was on policy-holders to take preventative action. In what might be considered a pre-emptive strike, the ABI drafted new policies to protect the insurance industry from large payouts because of Y2K. The policies, which started appearing in September 1998, prohibited claims for business interruption and property damage caused by Y2K. In product and professional indemnity, the policies made it impossible to claim back Y2K-related financial loss from legal liability and the costs of mounting a defence, if one were sued for instance (Adams, 1998). That noted, insurers did offer Y2K-related coverage but at a premium, and insurance companies typically insisted on audits of potential policy-holders' systems before such coverage could be agreed (Kelly, 1998).

There were still relatively few legal precedents relating directly to the IT field in general, and Y2K in particular. Academics continued to debate professional standards, responsibility and competence among IT professionals (Rowland, 1999). In the face of the uncertainty some law firms saw Y2K as a potential 'cash-cow', and perhaps not surprisingly there was a growth within firms of Y2K specialists (Hammam, 1998). The uncertainty and liability drove many state governments to draft laws to protect themselves. At least seven states considered such legislation—Nevada, California, Indiana, New Hampshire, South Carolina, Virginia and Washington.

In the United Kingdom, there was no clear view about the potential success of Y2K lawsuits. While Y2K received scant attention in academic communities, the UK-based *Journal of Information, Law and Technology* ran a special Y2K edition in June 1999. Howells (1999) concludes that non-Y2K compliant products put into circulation after 1 January 1990 causing personal injury, death or property damage would satisfy the requirements for a claim under the European Economic Community (EEC) Directive on Product Liability.[9] He bases the starting date on the date on which Y2K 'was known' though he cites no specific evidence to support his claim about that date. Peysner (1999), on the other hand, predicted that there would be little Y2K-related litigation in the United Kingdom in the run-up to 1 January 2000. He argues that pro-active, pre-event legal advice would tend to mitigate the effects in the commercial world. In the context of retail sales of defective software and/or hardware, the financial and procedural environment in the United Kingdom is unfavourable to mass litigation. Peysner sees too much risk and too few incentives for the lawyers, and concludes that high damages, risk free litigation, active lawyers and user-friendly procedures, conditions present in the United States but absent in the United Kingdom, must be in place before mass litigation occurs. Rather, he predicts a pressure to cooperate emerging in advance of the date-change, which is indeed, what we saw in the United Kingdom.

In the United Kingdom, negotiations were common as many of the formal regulating practices of the market were ineffective. There was no specific Y2K legislation in the United Kingdom. One private member's bill was introduced in the Spring 1997 but died as a result of the election call. When the new parliament convened it was decided there was insufficient time to introduce a new bill. Instead, CCTA recommended that organizations should move quickly to arbitration in the event of legal disputes. This point was reinforced by the Standing Committee on Science and Technology (CCTA, 1997d, 35, 36; Standing Committee on Science and Technology, Recommendations, 2). As a result, the only recourse open to government or businesses was the *Supply and Sale of Goods Act, 1994*, which allowed claimants to contest sellers of faulty goods (provided the goods were sold midway through 1993 or thereafter). That noted, the government was not always consistent with its own position. While being questioned by the Public Accounts Committee in June 1998, Robin Mountfield, permanent secretary at the Cabinet Office, acknowledged that the government was investigating the possibility of securing compensation from manufacturers of computer equipment that had to be modified as a result of the bug (Cane, 1998).

Perhaps more importantly, we see that the law is not merely part of the context but is part of the managerial response, and therefore, the law, like other managerial tools, is subject to change, especially in the face of a crisis. For instance, Peysner's observations run aground when we consider how few lawsuits materialized in the United States. Y2K legislation is one reason why there were so few lawsuits. In the United States, the Congress introduced five pieces of legislation concerning Y2K. The executive threatened to veto them several times but President Clinton eventually signed into law on 19 October 1998, *The Year 2000 Information and Readiness Disclosure Act*, and on 20 July 1999, *The Y2K Act*. Among other things, they imposed caps on punitive damages at $250,000 for companies with fewer than 50 employees; tighter standards of proof of liability, including the exclusion of liability if an organization offered advice 'in good faith'; and, in the Y2K Act, a 90-day waiting period after any problem appeared during which the sued company would be allowed to fix the problem. The Act also suspended 'joint and several liability', which normally makes any one defendant liable for the entire judgement. The US Government said the legislation was intended to reduce frivolous lawsuits.

Ultimately, many private organizations in the United States and the United Kingdom, like the UK government, opted for a much more cooperative approach, such as arbitration or joining sector-level information-sharing bodies rather than the more confrontational environment of a court-challenge (Jacobs, 1998).

The Acts were not the only reason for the few court challenges in the United States, however. Even as the legislation was appearing, little had materialized in the US courts. According to a study by Pricewaterhouse Coopers and cited by the Senate, by 30 June 1999, there had only been 74 Y2K lawsuits filed in the United States, of which 45 (61 per cent) were unique cases; 65 per cent were non-compliant product cases; 13 per cent were class-action shareholder suits; 4 per cent were insurance claims; 2 per cent were contractual disputes; 9 per cent concerned remediation efforts; and 7 per cent concerned false or misleading disclosure (US Senate Special Committee on the Year 2000 Technology Problem, 1999b, 158). Ultimately, fewer than 100 Y2K-related federal or state lawsuits were ever filed (Thibodeau, 2000).

One reason for a more cooperative stance is simply that organizations' IT suppliers were invariably the ones upon which the organizations depended most for assistance and advice in the run-up to the date change-over. No one really knew the systems as well as the IT service providers and therefore, organizations could either challenge the service providers in a risky legal environment or work with them to find solutions to their Y2K problems. Most opted for the latter. Indeed, Barclays Bank noted, 'relying on legal remedies to address the problem is illusory' (Standing Committee on Science and Technology, 3, 7).

That noted, there were still some sizeable legal challenges, particularly towards insurance companies, and the Senate noted its concern. For example, GTE, Xerox and Unysis sued insurance companies with whom they had policies to recover the money they spent to repair and test their systems. GTE alone sued for $400 million (Associated Press, 1999; Lohse, 1999b). The companies invoked the obscure 'sue and labor' clause that allows a company to reclaim money it spends from an insurer if the expenses ultimately saved the insurer money. The Acts did not directly address insurance issues.

## The market failure hypothesis

In the MFH, regulatory regime content reflects the inherent nature of each risk, and specifically the extent to which it is feasible for markets, including insurance or the law of tort, to operate as regulators of risk. Regulatory size and regime structure reflect the scale of the relevant hazards (Hood *et al.*, 2001, 70).

As a means of testing the explanatory power of the MFH, Hood *et al.* select two costs that can lead markets or tort law processes to fail in handling risks: information costs and opt-out costs. Information costs are faced by individuals in their efforts to assess the level or type of risk to which they are exposed. From a MFH perspective, Hood *et al.* expect the size of regulatory regime content to be larger for high cost cases than low cost ones because individuals would be more likely to resist expensive information-gathering activities unless pressured by government intervention (73). 'Opting-out' costs are costs individuals incur to avoid risk exposure through, among other things, civil law processes or insurance. The cost of individually opting-out of a hazard can be considered in absolute terms, but it can also be considered relative to a collective opt-out strategy (73).

If the market failure approach to risk regulation is followed, regulatory size is substantial only for risks where opt-out costs and information costs are high, and only for the specific control component that is affected by high costs. Conversely, if both information and opt-out costs are low, the market failure approach leads us to expect regulatory size to be small. If information costs are high but opt-out costs are low, market

|  |  | Cost of obtaining information on exposure to risk | |
|---|---|---|---|
|  |  | Low | High |
| Costs of opting-out of exposure to risk by market or contractual means | Low | Minimal regulation | Regime content high on regulatory size for information-gathering, with behaviour modification through information dissemination |
|  | High | Regime content high on regulatory size for behaviour modification | Maximal regulation |

*Figure 4.2*  The logic of a market failure approach to regulatory size.
Source: Hood *et al.*, 2001. Reproduced with permission.

failure logic suggests regulatory size is high for information-gathering but low for behaviour modification. If information costs are low but opt-out costs are substantial, regulatory size is low for information-gathering but high for behaviour modification. Figure 4.2 summarizes Hood *et al.* expectations of an approach to regulation dictated by the logic of market failure.

## The market response

Many organizations—within and outwith government—did spend a considerable amount on Y2K. The big spenders were almost always US companies. International Data Corporation (IDC) and the US Department of Commerce estimated the cost to the US economy to be approximately $100 billion (Gantz, 1997). To put the cost in comparative perspective, Figure 4.3 compares the US Government departments that spent the most with the companies from Wall Street that spent the most and some other organizations highlighted by the Senate's Y2K Committee. The US and UK governments' expenditure on Y2K at $8 billion and £400 million respectively are not completely out of line with other large organizations. Indeed, only the US Treasury stands out as having spent considerably more than the rest.

Companies like the ones noted below started early; they were wise to do so. Ultimately a longer-term Y2K plan that started in the mid-1990s

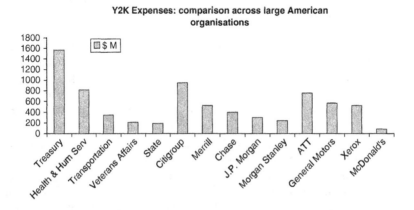

*Figure 4.3* Total Y2K expenditure (Selected Organizations)
*Source*: OMB, 1999; Smith and Buckman, 1999; US Senate Special Committee on Year 2000 Technology Problem, 1999b. Excludes the Department of Defence.

tended to be more cost-effective than a short term one that started in the late 1990s. The technology industry was growing exponentially: it was generating new products and services that presented companies with the opportunity to achieve efficiencies, increase reliability and expand into new markets, such as retiring mainframes for mini computers or client–server environments; going online; developing e-commerce options; and simply expanding the use of IT.[10] Moreover, as noted above, many organizations did not have reliable inventory lists; nor did they have contingency plans or sufficient IT security, an issue that was growing in importance given the increasing incidents of hackers (Grande, 2000; Smith, 2000; also benefits noted in several interviews). All three were concerns of some note but varied in terms of their priority. Because Y2K was often a top-down, 'every system must be checked' approach, Y2K offered organizations the opportunity to gain considerable advantage by simultaneously advancing in some of these other issues. Such an approach, however, did favour larger organizations that were cash-rich and had large IT budgets and long range strategies.

The CAA is one such example, although it started later than most large private organizations. Much of its Y2K work was carried out simultaneously with other work programmes. For instance, the CAA was upgrading from Windows 95 to Windows 2000 at the time of Y2K-related work. The decision to upgrade was not based solely on Y2K. Another example—the Engineer Licensing System—was upgraded to meet Joint Aviation Authority (JAA) regulations and was made Y2K compliant at the same time. While these projects would have gone ahead anyway, according to the CAA, non-Y2K compliance was always noted in the business proposals and in some cases may have helped expedite the process (INT 23).

Indeed, this is one of the reasons why Y2K costs must be interpreted with caution. By late 1999 most large organizations and banks in particular wanted to demonstrate that they had taken Y2K seriously to avoid overreaction by an anxious public. One way to do this was to indicate how much companies spent on the problem—the bigger the better. Nevertheless, many interview subjects, from the IT industry as well as the public sector, confirmed that much more was accomplished with this money than simply rectifying the Y2K problem. In any event, specifying exactly what Y2K cost is difficult for other reasons. For example, most large organizations have IT replacement policies that state how often systems are replaced, and therefore, the cost of Y2K would simply be the cost of *accelerating* the purchase from whatever year the organization planned to make the purchase to the year the organization made the purchase to avoid Y2K-related problems. In fact, the NAO discovered that the counting of Y2K costs was not done consistently across departments and therefore the final numbers are unreliable (INT 60). In the United States, where Y2K proposals were funded centrally, interview subjects noted that budget proposals were re-written in order to include Y2K aspects to the proposals to ensure the proposals were funded. But in effect, the Y2K aspect was peripheral to the request as far as the department was concerned (INT 26). In short, most Y2K costs are overstated and a final 'bottom line' remains elusive.

In any event, cash-wielding organizations spending such apparently large sums on their Y2K efforts were more the exception than the rule. Most organizations started well after the mid-1990s. Indeed, the market in general was slow to investigate Y2K problems. The typical Y2K publication attributes the relative lateness in starting Y2K projects (*c.* 1997–1998) to programmers believing that most systems that were developed in the 1970s and 1980s would be replaced before the year 2000. There was also hope of a 'silver bullet' solution coming along to save the day that never did appear. Finally, there is a sort of chicken and egg argument: as long as there was no demand for Y2K compliance— and there was relatively little in the early and mid-1990s—then the IT industry was not particularly driven to provide Y2K-related services.

This uncertainty created significant problems for larger organizations: they depended significantly on SMEs in their supply chains. After all, what if these SMEs did not know or believe that their organizations had Y2K vulnerabilities and they were prepared to take the *chance* that there would be no problem? Avoiding the costs associated with expensive audits or fixes were also pressures acting on organizations, particularly SMEs working on smaller margins, with little or no specific IT budget. This gap opened up a space for government intervention.

### Government intervention: market failure or market enforcer?

Governments controlled their internal process by bureaucratic mechanisms paying scant attention to market signals. It was a *precautionary approach*[11] aimed at eliminating the risk. The OMB and the CO's decision to oversee all systems from all agencies precluded a measured or balanced approach, particularly in the United States, which funded the plan centrally. Both governments paid little heed to the technical nature of the risk (for example, date functionality in systems). Few interview subjects note any attempt at cost benefit analyses or probabilistic risk assessments, for example.[12] There were also inconsistent attempts across

**Validation of Y2K process**

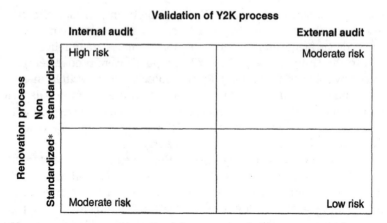

*Figure 4.4*   How the governments understood risk in the run-up to Y2K
*Standardized process: inventory, fix, test and contingency plan.

government at prioritization of systems. Departments and agencies may have identified 'mission critical systems' but OMB/CO collected information on all systems and (indirectly) pressured departments to check and fix *every* system. And the process could only be truly confirmed by external validation in the form of an outside auditor. The 'risk lens' through which the governments understood the problem can be depicted as in Figure 4.4

There are many examples of both governments over doing it. BLS staff at the International Price Program had upgraded their systems and felt confident that they showed no vulnerability but they were forced to go through an expensive overhaul of systems that took professional IT staff over a year with little reference to the opportunity costs of such a decision (INT 41). One senior staff member at the FAA estimated that the FAA had only to replace about 20 per cent of its systems but in fact fixed all of them because of pressure from the executive, the political staff at the FAA and the media (INT 29). In all cases, the size of the reaction was even greater when one considers that the standard Y2K process included two layers of redundancy-external audits *and* contingency plans.

One might claim that the markets are of little import to government departments and agencies because they are not market-driven. While this may in fact be the case, advocates of New Public Management (Hood, 1991) and Reinventing Government (Osborne and Gaebler, 1992) trends—arguably in their hey-day in the mid- to late 1990s—would suggest that government departments and agencies ought to have been sensitive to price signals and market pressures in the run-up to Y2K. Certainly concerns such as keeping costs down and revenues up are central to many custodians of public sector budgets in the short and medium term. What Y2K demonstrates quite convincingly, however, is that when the chips were on the table, the bureaucracies reverted back to top-down, command and control structures.

Intervention in the market was potentially complicated. To return to Hood *et al.'s* anticipated response, people had few exit options. In fact, the distinctions drawn by Hood *et al.*, opting-out of risk/obtaining information about risk, were not mutually exclusive in the Y2K context. In many respects, particularly in the early period, the bug was understood as pervasive and unpredictable, and obtaining reliable information or opting-out of the risk remained elusive. Information gathering and risk reduction came hand in hand as people trawled through systems and occasionally fixed date fields as they did so. This is particularly so in larger organizations. No one was prepared to say with certainty that systems would not fail until complete inventories were conducted, the date-fields were fixed and the audits were carried out. And even then they could not be certain about those suppliers upon whom they relied but whose systems' Y2K compliance was uncertain.

We should therefore expect to see 'maximum' regulatory response, and indeed, we see considerable interventions by both governments. The governments increased the profile of the problem and brought people together to share ideas and assure one another that they were working on the problem. They also increased transparency by reporting sector-wide Y2K compliance and encouraged sector-driven results and standards. As noted, in the United Kingdom the government went so far as to declare a UK-wide strategy. In this sense, the UK government in particular played a powerful information-gathering and enforcement role.

The reliance of the supply chains was always problematic, however. As the SMEs in particular played an important role in supplying the large companies they too had to be reached. Again the government played a communications/information-disseminating and gathering role. They polled SMEs often and they ran communications strategies aimed at SMEs and micro-businesses.

Despite its seemingly forceful intervention in the market, government communiqués were always more convincing than the reality on the ground. In the case of aviation, a high safety standard can drive up its economic value. One might argue, therefore, that 'over-managing' was a sound economic decision. This is no doubt true, but the pressure the governments applied to their own operations was not applied with the same *rigueur* to those organizations outside the governments' immediate control. The FAA, for example, was never able to secure guarantees from all those it licensed that they were compliant, nor did the FAA ever intend to do so, despite pressure from Congress (INT 45). In the case of the CAA, it turned a blind eye to the fact that 22 organizations did not submit compliance statements and further, it allowed one organization in the 'Business as Usual' programme to drop out of the programme without penalty (Action, 2000, 1999). In short, while governments were effective at information-gathering they were rather less successful behaviour modification through direct intervention.

Relatedly, the timing of government intervention was not always optimal. Advocates of the Y2K Acts, for instance, note that the legislation allowed organizations to share information so that they could learn from each other so that organizations would not have to check the same information. This is no doubt true but in many respects by late 1998 the legislation seemed a little late, and perhaps largely symbolic. The organizations that had started in the mid-1990s or even 1998 for that matter often duplicated what had already been done by similar organizations, each one investigating similar equipment and often failing to find Y2K-related problems. Indeed, information-sharing earlier in the process might have been more fruitful. This US government approach was only optimal for IT organizations selling standardized Y2K testing methods, wishing to keep demand for their product as high as possible for as long as possible. Once the IT industry was confronted with the possibility of lawsuits then the Y2K legislation became useful for it.

What is perhaps most intriguing is not how the governments regulated industry in the run-up to Y2K but rather how government, and the US government in particular, withdrew from the market. The governments did not pass Y2K compliance legislation; they were silent as the insurance industry—the repository of contingency funds—excused itself from Y2K; the governments also loosened the legal constraints on the IT industry. In the United Kingdom it did so by actively encouraging arbitration and by not legislating on the matter. In the United States, the government tried to change arrangements to make it more like the United Kingdom— it removed temporarily people's right to sue and limited the damages they could seek. In fact, the governments let the market run its course: negotiations; self-interest; survival; market demand for Y2K compliance; and continuity of service. These were the pressures that applied. In many respects, one was on one's own.

Was that reasonable? Certainly it has a Darwinian ring to it; it also suggests that government has limited control over a rather fragile infrastructure. One argument suggests that the belief in the complexity and interdependence of the systems with the short time frame changed the conventional risk calculations. When the survival of the firm is at stake, risk can no longer be described as the product of probability and expected monetary losses. A more appropriate description in these cases can be attempted in terms of cardinal utilities. When the firm is in danger of massive operational failure in the short term, conventional, long-term risk assessments do not necessarily hold. A logic driven by short-term survival rather than long term return on investment emerges. So it was with the government. The infrastructure was under threat and government opted for 'survival,' so to speak. It did so by creating an exhaustive and largely inflexible process within government. In the market it helped to create sector-level pressure to ensure organizations moved towards compliance.

This strategy has problems of its own. Some government actions, and indeed inactions, allowed other market failures to surface. For instance, when the government pressures organizations to seek information, this creates an incentive for others to enter the market and provide that service. With a limited time-frame like Y2K, this pressure drove down the quality of information. Similarly, when the governments chose not to regulate formally behaviour modification among SMEs, larger industries filled the gap by pressuring SMEs in their supply chain to undergo expensive Y2K audits. Some of these SMEs felt they had no Y2K vulnerabilities but had limited ability to push back among the industry giants that frequently dominate the sector and the supply chain. The governments' reluctance to act created a vacuum, which powerful companies filled, but it was not necessarily a cost-effective response at an aggregate level.

What might have helped the governments' information-gathering/disseminating strategies is if they had started earlier than they did, which would have allowed them, perhaps, to distinguish better between reliable and unreliable information. Yet this approach has its own challenges. The earlier the government intervenes the less guidance it has from the market about what intervention might be appropriate. If the governments had pressured all organizations to investigate Y2K-related vulnerabilities, then they would have run the risk of repeating the overreaction that occurred, except for longer. They would also be encouraging organizations to absorb a small amount of Y2K-related costs every year in the run up to Y2K, dismissing the possibility that a 'silver-bullet' solution might present itself closer to the date-change. Alternatively the governments might have targeted particular industries that seemed vulnerable. But in this case would the governments' strategy be to target those industries that seemed most effected by the 'technical nature of the risk' (for example, banking) or would they also pressure sectors that were more likely to be effected by people's perception of the risk according to psychometrics (for example, aviation; nuclear facilities)?

Indeed, managing public perception of the risk was every bit a part of managing the Y2K problem. But managing public perception was not merely a matter of ensuring everyone remained calm. In fact the opposite is true; crises can generate action and it was therefore a tool that governments could—and did—use to enact their preferred strategy, a point explored in the next chapter.

# 5
# Opinion-Responsive Hypothesis

Contrary to popular depictions of reactions to Y2K in which the public hoarded basic supplies because of a considerable fear of an Armageddon-like disaster, such as those found in Michael Moore's documentary, *Bowling for Columbine* (2003), the real public reaction to Y2K was in fact much more understated. That is, in the face of potential operational shut-downs in which even some modest planning could have saved considerable inconvenience, the public tended not to deviate from its set pattern and routines.[1] The anxiety levels expressed through public opinion polls and most media coverage dropped between 1998 and 1999. The drop in anxiety levels can partly be attributed to robust responses to Y2K from government and industry, which included *inter alia* much more sophisticated communications plans. And indeed, the methods applied for learning business and public opinion likely led to poll results that exaggerated people's anxieties and businesses' vulnerabilities.

Y2K complicates the opinion-responsive hypothesis because public opinion was something governments felt had to be monitored (i) to ensure people were sufficiently aware of the bug such that they would fix their own systems; and (ii) to ensure that people did not panic and as a consequence jeopardize national stability through the hoarding of limited supplies, for instance. As a result the CO and EOP had a close eye on public opinion throughout the process but their actions suggest they tried to respond to it by shaping it. At first, their strategies suggest they tried to align public perceptions with the pre-eminent positions of experts, which advocated a robust response. Latterly, however, governments 'talked down' the risk in the hope of reducing public anxiety and the threats that it could potentially cause. In short, elements of style may have mirrored public opinion but size and structure remained stable from mid-1998. Ironically, it seems more the case that content shaped context, that is, government management deliberately tried to provoke a certain reaction within the public in order to meet its objectives.

While the governments may have been content with the timely rising and falling of public anxiety on the issue, however, it is debatable how much impact their interventions actually had. Government reacts with public opinion in a complex, overlapping, interactive process rather than in a dichotomous one that separates government actions and public opinion by a clearly drawn line.

This chapter focuses primarily on popular opinion polls[2] and the broadsheet media and attempts to explain government action in the light of these two influences. Note, however, two amendments that I have applied to the Hood *et al.* approach. First, while Hood *et al.* looked only at one mainstream broadsheet and one tabloid, I will consider a greater number of sources, including selected examples of the IT trade press and the financial broadsheets. The journalists and the readership of these two sources often had a significant stake and experience in Y2K management. The research findings would be conspicuous by their absence.

Second, while Hood *et al.* examine volume of coverage alone (that is, number of articles published) it is debatable that such a method provides an adequate analysis of the Y2K media coverage. As the analysis will show, volume as well as tone and content provides insights that volume alone cannot provide.

## Media analysis

One method Hood *et al.* use to determine public opinion is to count the number of articles on the subjects in question in two different newspapers, a tabloid and a broadsheet. They note that their analysis does not assume that high circulation newspapers reflect public opinion, but they do assume that it reflects 'the flavour of the public debate, not least because opinion leaders read such sources' (93). Hood *et al.* draw on Gaskell *et al.* (1999) for this analysis. In Gaskell *et al.*'s analysis, they conclude that increasing amounts of coverage of technological controversies are associated with negative public perceptions (385), or what is referred to as Quantity of Coverage Theory (Leahy and Mazur, 1980). While Hood *et al.* do not define what they mean when they refer to the 'flavour of public debate', we might assume from Gaskell that the greater the volume of coverage the more negative the perceptions.

To start, I will make a brief note on the IT press and the tabloid/ 'populist' press by way of acknowledging their coverage and situating it in relation to that found in the broadsheets, which, as stated, together with the popular opinion polls, will be the source of most of the analysis of this chapter.

## IT press

Between 1 January 1997 and 31 December 2000, *Computerworld*, a leading IT trade magazine based in the United States, printed 422 stories on Y2K.[3] Like the broadsheet press, the coverage grew throughout 1998 and 1999, though it peaked a little bit earlier than the other, in the third quarter of 1999. The tone of the headlines was more frequently alarming than reassuring, by a ratio of approximately 4:3, and more consistently so. There did not seem to be a momentum at different times for alarming headlines over reassuring ones, or *vice versa*, as occurred in the other American papers.

In a review of articles that were published in the UK-based *Government Computing* between October 1998 and December 2000, I identified 21 articles about Y2K,[4] of which 38 per cent had alarming headlines and 24 per cent had reassuring headlines.

The tone of the coverage in the IT press was similar to that found in the financial broadsheets. While on the one hand one might have expected a somewhat more moderate or balanced approach from the IT press, especially as IT specialists tended to be more confident than others that Y2K would not result in disaster as 1 January 2000 approached (Gutteling and Kuttschreuter, 2002), in fact, the sources that were being used in *The Financial Times* and *The Wall Street Journal* were by and large the same ones that were being used by the IT press. In the early days in particular (*c.* 1997), there were not that many Y2K specialists around and the few that existed were referred to by all publications.

## Tabloid press/popular press

Comparing the tabloid/popular press to broadsheets is not as straight-forward, because the two papers in the popular press show starkly different trends. *The Sun* hardly touched Y2K, printing 15 stories in total, whereas the *USA Today* printed 157 stories, covering the story even more often than the mainstream broadsheets. Figure 5.1 summarizes the different trends in coverage by each tabloid. The *USA Today*'s interest is perhaps not surprising given the high level of public anxiety on the issue from late 1998 and given that the story leant itself

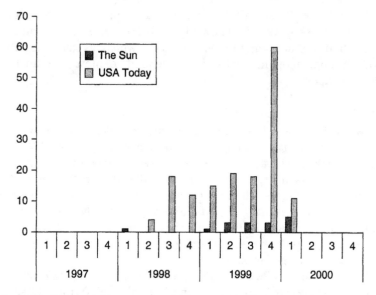

*Figure 5.1*   Volume of coverage in *The Sun* and *USA Today*, sorted by newspaper and by quarter, 1997–2000

to dramatic headlines, such as the threats posed to Russia's nuclear arsenal as a result of Y2K. What is perhaps more surprising is *The Sun*'s lack of interest. One senior journalist (not an employee of *The Sun*) speculated that sometimes newspapers are reluctant to carry a story that they missed at the outset (INT 31). Certainly it is easy to see why *The Sun* would have missed the story—in 1997 it was largely considered a business one, a small and relatively insignificant section to *The Sun*. It would also perhaps help to explain why *The Sun* took such a cynical view of the story in the few articles it did run after 1 January 2000 (Table 5.1).[5]

In sum, as the reader will come to see, the *USA Today* coverage is quite similar to that found in the broadsheet coverage. *The Sun*, however, is the outlier. Despite Y2K's potential to provide rather sensational headlines, *The Sun* ignored it.

### Broadsheets

The remainder of the analysis is derived from the broadsheets and public opinion polls. The following newspapers have been selected as an overview of print news coverage of Y2K: *The Wall Street Journal* (WSJ),

*Table 5.1* Tone of tabloid/populist media coverage: percentage of alarming headlines to reassuring headlines, sorted by newspaper, 1997–2000

| Newspaper | Alarming | Reassuring |
| --- | --- | --- |
| USA today | 32% | 31% |
| Sun | 33% (5) | 27% (4) |

*The New York Times* (NYT), *The Financial Times* (FT) and *The Times* of London (LT).[6] The papers were selected because they are among the most read papers in their respective countries and because they represent two 'types' of newspapers, financial and mainstream, which attracted different but overlapping audiences in the run up to Y2K.

*Volume of broadsheet coverage*

The financial newspapers covered the story more than the mainstream papers did. The FT's total coverage, 218 articles, started early in 1997 and continued, albeit with some peaks and troughs, throughout 1997 and 1998. Interest peaked in the FT, like every other broadsheet in this study, in the fourth quarter 1999. The WSJ covered the story the most, with 262 articles. Its coverage of Y2K started later than that of the FT (early 1998) but grew rapidly, each quarter showing greater volume, with the exception of a slight reduction in the second quarter of 1999.

With respect to the mainstream papers, the LT printed 35 per cent more stories than the NYT, producing more stories than the NYT in each quarter, with the exception of the second quarter 1998 and the fourth quarter 1999. The NYT demonstrated slightly more interest in Y2K after 1 January 2000. Overall, however, the pattern of coverage in both papers was similar in that, comparatively, the broadsheets showed moderate and consistent interest in the story throughout.

When the coverage is sorted by country, the United Kingdom started its coverage earlier and published relatively consistently on the topic from the fourth quarter 1997 onwards. There was a noticeable bump, however, in the fourth quarter of 1999. Its early coverage can be attributed largely to the coverage in the FT. The US coverage, on the other hand, started later but grew quickly. Figures 5.2 and 5.3 summarize the results by newspaper type and by country, respectively.

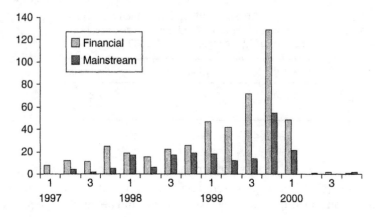

*Figure 5.2*  Volume of broadsheet coverage sorted by newspaper type and by quarter, 1997–2000

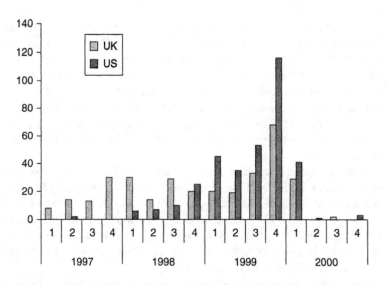

*Figure 5.3*  Volume of broadsheet coverage sorted by country and by quarter, 1997–2000

### Tone of the broadsheets

On a national level, the headlines in the UK press were more alarming than the headlines in the American press on the whole. Forty-one per cent of UK stories had alarming headlines whereas only 32 per cent of

American stories had alarming headlines. UK newspapers were also only half as likely to publish a reassuring headline as their US counterparts.

However, tone varied considerably over time. The US coverage started in 1998 with 75 per cent of Y2K stories with alarming headlines, a level never reached in the United Kingdom. From the start of the coverage in the first quarter 1998 to the second quarter 2000, the US newspapers became less and less likely to use alarming headlines. From the first quarter 1998 to the first quarter 1999 there was a downward trend, peaking at 75 per cent at the outset and dropping to 24 per cent. They plateaued in 1999, staying within 24 per cent and 31 per cent. The UK coverage, in contrast, was much more volatile. While proportions of alarming headlines also went down overall, the drop was only slight, and there was much variation in the intervening period. Of the 11 quarters between the second quarter 1997 and the fourth quarter 1999, only on two occasions was a movement in one direction followed by a similar trend the following quarter. Figure 5.4 summarizes the percentage of articles that used alarming headlines, by country. Table 5.2 summarizes the percentage of 'alarming' and 'reassuring' headlines, by newspaper.

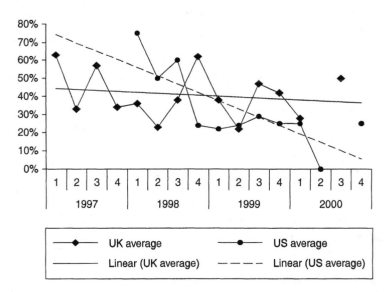

*Figure 5.4* Tone of broadsheet coverage: percentage of articles with 'alarming' headlines sorted by country and by quarter, 1997–2000

*Table 5.2* Tone of broadsheet coverage: percentage of alarming headlines and reassuring headlines sorted by newspaper, 1997–2000

| Newspaper | Alarming | Reassuring |
|-----------|----------|------------|
| FT | 52% | 13% |
| LT | 25% | 20% |
| NYT | 27% | 29% |
| WSJ | 39% | 37% |

## Sources and references contained within the broadsheet coverage

'Sources' refers to events, publications and/or people that journalists use in a story. Stories can have numerous sources. Table 5.3 summarizes the percentage of articles in the four broadsheets examined that employed the noted sources. 'References' refers to institutions, sectors, jurisdictions and concepts to which articles refer. Articles can have more than one reference. Table 5.4 summarizes the percentage of articles in the four broadsheets examined that contained the noted references. To illustrate the difference between sources and references, note that, for example, IT consultants were frequently sources for stories because journalists would interview them on numerous aspects of Y2K. The articles rarely referred to the IT industry; more frequently they referred to the cost of compliance; or the readiness of other countries; or the readiness of the government. The sources and references identified here were selected either because they were common references or sources in the articles, or because they were conspicuous by their absence. Note, 'Critical Sectors' refers to important sectors for the infrastructure, such as those participating in the UK/NIF or the US/WG. 'Speculative Estimates' refers to broad, macro-level forecasts and predictions about the outcome of Y2K. They were stated as probabilities in statistical form (for example, 50 per cent chance of a major failure in the power sector). 'Other jurisdictions' refers to references to governments other than the US Federal Government by American newspapers or references to governments other than the UK Central Government by UK papers. They can include sub- or supra-level governments as well as other national governments. The data in Tables 5.3 and 5.4 will be referred to in the next section, which summarizes the broadsheet coverage.

## Summary and analysis of the broadsheet coverage

The 1997/early 1998 coverage oscillated between alarming and reassuring, though it was mainly the former, due largely to the uncertainty

*Table 5.3* Selected sources within the broadsheet coverage

| UK | | US | |
|---|---|---|---|
| IT industry | 24% | Critical sectors | 49% |
| Critical sectors | 23% | IT industry | 28% |
| National government | 22% | National government | 27% |
| Industry associations | 13% | Regulators | 20% |
| Regulators | 12% | Industry associations | 16% |
| National legislature | 6% | National legislature | 12% |
| Academic | 2% | Academic | 5% |
| Popular polls | 0 | Popular polls | 3% |

*Table 5.4* Selected references within the broadsheet coverage

| | UK | US |
|---|---|---|
| Other jurisdictions | 36% | 32% |
| Departments / agencies | 21% | 20% |
| Cost of compliance | 21% | 19% |
| SMEs | 8% | 8% |
| Actual Y2K incident | 6% | 8% |
| Speculative estimate | 6% | 5% |
| Benefits of Y2K | 5% | 3% |
| Y2K-related terrorism | 1% | 3% |

surrounding the scope and magnitude of the problem. In this particular period, newspaper articles rarely included a detailed explanation of what the Y2K computer bug was and how it could disrupt operations. If the nature of the Y2K bug was described at all, it was usually in one short paragraph, which referred to a 'glitch' resulting from many computers' inabilities to read dates. This is not surprising since detailed explanations would have quickly become boring, but the frequently brief explanation had other implications. It meant many people had only a superficial understanding of the problem. In a short time 'Y2K' became a generalized concept that implied significant risk of operational failures around the millennium. Companies' Y2K vulnerabilities were described in broad-brush strokes according to their dependence on IT and/or external suppliers and the consequences of 'worst-case scenario' systems failure in critical sectors (for example, power failures, grounded aviation, food and water shortage, inoperable health care and emergency services).

Y2K specialists were relatively few and they got quite a bit of coverage. As noted, about one-quarter of all articles in both countries used an IT source. The coverage that refers to Y2K specialists and consultants was almost completely alarming. Six per cent and five per cent of articles in the UK and US print media, respectively, contained 'speculative data', which was usually provided by a source from the IT industry and largely turned out to be exaggerated as noted in the previous chapter. Edward Yardeni from Deutsche Bank, for instance, famously predicted that there was a 70 per cent chance of a world recession as a result of Y2K (1998). At the same time, there was little central coordination in government, which often resulted in individual departments' communications teams being put on the defensive. The sum of these circumstances leant themselves to rather spectacular and alarming headlines.

But it was not just the media that had adopted an alarming tone. The House of Commons and the Congress were issuing their own reports, which were equally alarming.[7] Moreover, some large, credible organizations started to declare how much they were spending on Y2K and the amounts were considerable. One of the first such declarations came from the Royal Bank of Scotland, which estimated Y2K would cost it £29 million (Graham, 1997). Some organizations speculated publicly on rather dramatic contingency plans, such as KLM claiming it might ground all its flights over the New Year period (*Financial Times*, 1998).

This period had considerable impact on the FAA in particular. One interview subject noted that communications on Y2K represented a significant challenge because the plan and the details were too complex to convey in the simplified terms necessary for dealing with the press. Despite his senior position, he was instructed not to deal with the media directly (INT 29). One interview subject from the communications department noted that the media's Y2K coverage was almost entirely negative and hysterical (INT 49). The CAA, in contrast, received comparatively little coverage. This dearth of commentary is particularly noteworthy because the CAA, while responsible for a much smaller operation than the FAA, did not seem to be ahead of the FAA in terms of Y2K progress nor did the CAA have a particularly good reputation for delivering large IT projects on time (INT 63; Leake, 1998). The difference is ironic and can partly be attributed to the difference between the two legislatures: the Congress was much more vocal and got much more coverage than the House of Commons, and indeed, the FAA was one of Congress's favourite Y2K targets. The BLS and ONS received no direct coverage at all (INT 29) but that does not necessarily suggest that they were not affected by media coverage generally. The dramatic claims made by the media (and Congress) permeated government as a whole. To an extent, the notion of 'planes falling from the sky'—frequently cited at the beginning of popular references to Y2K—served as an availability heuristic;[8] it portrayed the potential consequences of Y2K in the most dramatic, albeit highly unlikely, terms and thereby helped to establish a serious tone that influenced Y2K projects across government as well as large private industries. As noted in the previous chapter, there were few nay-sayers and they got little coverage. Note, also, that the press rarely referred to academics.[9]

The financial papers were particularly interested in the story because by mid-1998 almost every large corporation had a Y2K plan, and in the United States the Securities Exchange Commission (SEC) directed public companies to disclose information about their plans. Companies claimed to spend anywhere from millions to tens of millions to hundreds of millions of dollars on it. The nine Wall Street firms that claimed to spend the most on Y2K, for example, reported that they spent $2.8 billion collectively (Smith and Buckman, 1999). Twenty-one per cent of UK articles and 19 per cent of US articles mentioned the cost of Y2K compliance, but in general spending large sums on compliance was rarely reported sceptically or cynically. The accuracy of the costing was never questioned. In almost all cases, the cost was reported as a way of showing the magnitude of the perceived problem. There was also a considerable amount of coverage given to financial analysts speculating on the Y2K readiness of companies/sectors. In this latter category, the difficulty in obtaining reliable information led to some highly suspect claims. Similarly, there was a professional consensus that financial systems were vulnerable to Y2K-related problems because most financial transactions are date-dependent. As financial newspapers, this vulnerability gave them reason to cover the story even more. Moreover, in 1997 and 1998, when the insurance industry seemed at risk of receiving numerous Y2K-related claims, the financial papers covered the story considerably.

From early 1999 many companies and government departments and agencies had large Y2K operations in place, and they were starting to see less alarming results. Equally important, as Y2K rose in the popular consciousness, many organizations, including Action 2000 and the President's Council on Year 2000 Conversion, were discovering better ways to communicate Y2K compliance to the media and to the public. In the main, however, the tone of the coverage became less alarming from early 1999 onwards as companies trawled through their IT and embedded systems, tested those systems and checked with suppliers to ensure that the suppliers were Y2K compliant. Companies also started developing contingency plans and then communicated their level of compliance through the press, shareholder meetings, annual reports, government fora and news releases. But there were other indicators. Some key dates were understood to be precursors to Y2K mayhem, such as the start of the new fiscal year in 1999/2000 or 9 September 1999,[10] but they passed without incident. So, as Y2K coverage increased throughout 1999—it continued to be a good, eye-catching headline—alarming headlines decreased, particularly in the United States. In contrast, alarming headlines continued in the United Kingdom, as noted, due largely to the coverage in the FT. This latter day 'alarmism' can partly be attributed to the FT shifting its focus from domestic matters to countries that seemed to be more vulnerable.[11] Note that 36 per cent of UK stories, largely from the FT, referred to 'other jurisdictions', which for the most part meant other countries.

Again, the FAA provides a good illustration of this shift. Believing public and congressional confidence in aviation was at risk, the FAA became much more aggressive in 1999 with respect to public relations. As stated in Chapter 3, the FAA tested air traffic at an event open to the media; it set (ambitious) milestones publicly and reported on its progress publicly; staff invited Y2K consultants who had been publicly critical of the FAA to the FAA to discuss its approach; it emphasized full compliance in news releases; and it tried to get headline-grabbing positive news by announcing, for instance, that the Administrator of the FAA, Jane Garvey, would fly on a plane during the critical time change-over (INT 38).

Interestingly, only 6 per cent of articles in the United Kingdom and 8 per cent in the United States referred to 'actual' Y2K failures. In most articles 'failure' was only hypothetical. Nevertheless, given the lack of concrete evidence of 'actual' failures, it is little wonder that the tone of coverage, for the most part, was less alarming in 1999 than in 1998. Indeed, one congressional committee staff member noted it became increasingly difficult to get the NYT to cover the story because there was very little progress or anything new to report from one day to the next (INT 53). As the end of 1999 approached, there was even an increase in stories about the benefits of Y2K, such as the development of contingency plans, reliable inventories, supplier lists and international and inter-organizational cooperation. These benefits were largely true but emphasizing the positive was also a convenient way for government and industry to emphasize the 'good' in what was an extremely expensive risk management exercise, whose scope and cost were starting to look exaggerated by the end of 1999, particularly among non-IT people.

In sum, at the outset in 1997/1998, most press coverage focused on elite groups and institutions—government, critical sectors, regulators, trade associations and the IT industry. The tone of the press coverage was anxious during 1997/1998, but as the volume of coverage increased through 1998 and 1999, the alarming tone decreased. We can speculate with a degree confidence that the decrease can be attributed to the massive and exhaustive Y2K operations in place by 1999, the dearth of any *actual* Y2K-related failures and a greater emphasis by industry and government to communicate Y2K operations and compliance to the media and public.

## Public opinion polling: what were people saying?

Most polling results were drawn from surveys with senior managers, particularly during the early period. There was certainly no dearth of surveys: PA Consulting, Greenwich Meantime, International Data Corporation (IDC) and Capgemini conducted polls drawn from large samples regularly. The surveys were often, however, self-assessments drawn from a sample of people who did not necessarily appreciate the magnitude of the task, or else the polls were products of standardized templates that could not detect the subtleties of individual organizations' IT operations.

To illustrate the problem with self-assessments, consider a poll conducted at the end of 1996 by PA Consulting Group and the UK government's Taskforce 2000 of UK senior managers in the public and private sectors: it noted that two-thirds of the senior managers claimed to be 'partly aware of Y2K' and almost one-third were 'fully aware' (NAO, 1997, 9). In its commentary on the poll, however, while the NAO (1997) acknowledged that there was a high degree of awareness, it was less certain about what organizations were *doing* to fix the problem and when they would be finished. This uncertainty was troubling; anecdotal information from early experiences with the Y2K problem suggested it was a bigger problem than most organizations initially thought and they almost always ran over-budget and could not meet their deadlines.

The standardized surveys that tried to determine what organizations were doing had their own problems. For instance, surveys typically asked respondents questions such as 'Have you investigated the Y2K problem?' and 'Have you followed the following steps to fix the problem?' If the participant did not respond positively to both of these questions then the polling company would categorize the respondents as not being Y2K compliant (INT 58). These broad-brush surveys were designed for a macro-view of Y2K-readiness but did not get into the specifics of what was reasonable for each organization to achieve. An organization attempting to develop a measured response to Y2K that focused solely on priority systems, for example, would likely be categorized as 'not-compliant'. So, too, would an organization that had little date-functionality in its systems but had not yet carried out a detailed inventory of its systems. 'Not being Y2K compliant' meant one had not conducted extensive checking and testing; it did not mean an organization would experience systems failures on or around 1 January 2000. Hence, the survey results almost always exaggerated the risk of technical failure. Like many of the templates developed at the Cabinet Office and OMB, the only way an organization could seem ready in these survey results is if it had developed a very thorough, exhaustive response.

## Popular opinion polls

Few interview subjects at departments or agencies recall any kind of two-way dialogue between agency staff and the public. At most, the Y2K lead at the BLS recalled speaking at a couple of events for specialist audiences (INT 27). Quite contrary to initiatives such as the Citizen's Charter (UK) or Reinventing Government (US), in which public services were directed to be more 'customer'-focused, there seemed to be little formalized means to obtain feedback at the agency level from users. By and large, obtaining information directly from and giving information directly to the public was handled by the executives' lead offices, Action 2000 and The President's Council. That noted, CAA staff said that they were aware of ticket sales in the aviation industry in the run-up to 1 January, which incidentally had not changed significantly from the previous year (INT 22).

Neither was there much polling in the early stages. John Koskinen recalled that his team had conducted one poll of the American public but referred to it rarely (INT 54). In the United Kingdom, no interview subject recalled any public opinion polling at all. The absence of popular opinion polls contrasts with 'New' Labour trends and polling practices at the White House at the time. (See, for example, Murray and Howard, 2002, 545–8.) The reason may relate to the fact that, despite the frequent political rhetoric about IT, it was rarely seen as an issue that divided along party lines or for which there was a perceived political opportunity. John Koskinen noted in an interview that the Republican Senate Leader, Trent Lott, told him that elected Republicans would support the President's

Y2K initiative because ultimately, if there were any Y2K-related problems, the public would blame *both* the executive and the Congress. He felt voters would not distinguish between party lines (INT 54). Hence, once the President committed the government to EO 13073 and virtually predetermined an exhaustive approach much of the conventional politicking took a back seat. In short, with both parties on the same side of the debate, popular polling for political ends seemed less necessary.

'Popular' opinion polls started appearing in the press from mid-1998 onwards, which interestingly was after both governments had made decisions about the manner in which Y2K would be managed. In December 1998, a substantial number of Americans surveyed believed there would be systems failures in banking (63 per cent), air traffic (46 per cent), food and water (37 per cent) and emergency services (36 per cent). Indeed in January 1999 a Y2K survivalist guide hit number 70 on the Amazon.com Best Seller's List[12] (Miller, 1999). By December 1999, however, apparent anxiety levels dropped anywhere from one-third (food and water) to a half (banking). The December results were: banking, 34 per cent; air traffic, 27 per cent; food and water, 25 per cent; and emergency services, 22 per cent. Figure 5.5 summarizes the results of five polls taken between the end of 1998 and the end of 1999.

In the United Kingdom there was far less polling, and the results are not as easily compared across time as the US polls are. Gallup did ask people in the United Kingdom about Y2K at the beginning of 1999 and then again at the end of 1999. (See Figure 5.6 on p. 110.) While the two sets of questions are not exactly the same and therefore any interpretation is necessarily constrained by this variation, like the United States, there was a downward trend in anxiety levels in the United Kingdom. While the January 1999 poll results looked like something closer to 'panic' as over half of those surveyed described Y2K as a serious threat to services with a full one-third expressing personal safety concerns, by December 1999 only 8 per cent aligned themselves with such catastrophic language ('very worried'). Nevertheless, it bears noting that in December 1999 the middle option—'somewhat concerned'—attracted 41 per cent of the respondents. As with so much of the Y2K story, there were many fence-sitters, people who felt relatively confident but not absolutely certain that things would be alright.

### A complication: did Y2K cause people to alter their behaviour?

When the National Science Foundation asked Americans whether or not Y2K concerns would result in people modifying their behaviour, the results were different from the poll results cited above. They revealed

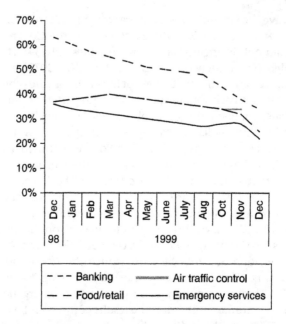

*Figure 5.5* Public opinion: percentage of Americans who said they think it is likely that these systems will fail as a result of the Y2K bug
*Source*: Jones, 1999; based on five data points: December 1998 and March, August, November and December 1999.

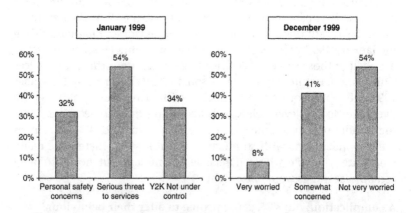

*Figure 5.6* UK public opinion: two surveys on how people felt about Y2K
*Source*: (1) 26 January 1999; ICL (MORI Polls); (2) Gallup Poll cited in *Coventry Evening Telegraph* 15 December 1999) (Gallup Organization and MORI websites in bibliography).

greater and sustained anxiety. In December 1998, 47 per cent said they or someone in their household had already decided or would decide to avoid air travel; 26 per cent said they would stockpile on food and water; 16 per cent said they would withdraw all of their money from the bank. In this case, there was no clear downward trend over the course of 1999. By December 1999, while 6 per cent said they would withdraw all money from the bank (down two-thirds), 51 per cent say they would avoid air travel (up 9 per cent from 1998) and 42 per cent said they would stockpile food and water (up two-thirds). There was no equivalent survey result for the United Kingdom. Figure 5.7 summarizes the poll results on changing behaviour.

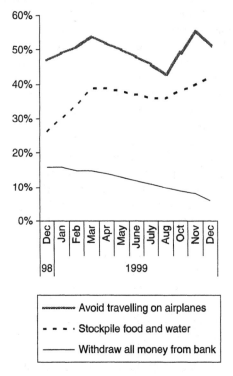

*Figure 5.7* Percentage of Americans who said what they or someone in their household had already done or probably would do to protect themselves against Y2K-related problems

*Source*: Jones, 1999; based on five data points: Dec 1998 and March, August, November and December 1999.

Despite this evidence of actual or anticipated behaviour modification, there is very little evidence to support a claim that people did actually change their behaviour. One article in the WSJ noted an increased demand for some medicines, guns, lip gloss, cigarettes, canned vegetables and wine among the US public (Starkman, 2000). At least three of these could be attributed to millennium celebrations. The other three (medicines, guns and canned vegetables) may be Y2K related. Yet, despite all polling results that indicated even moderate levels of anxiety in the United States, there was little evidence to suggest that either the US or UK population modified their behaviour to any significant degree (Zuckerman and Wolf, 1999). There may have been slight inflation due to Y2K-related purchases in the fourth quarter 1999, though this cannot be attributed conclusively to Y2K. The media reported anecdotal shifts in demand, but never enough to constitute a trend. Somewhat ironically, perhaps the only exception was the drop in demand for IT in the last quarter of 1999. Many organizations introduced an IT purchasing freeze; they did not want to introduce new bugs into their systems in the late stages of Y2K preparations (Kehoe, 1999). (This point will be developed further in the next chapter.)

There were, however, examples of the governments and key sectors preparing for behaviour modification that did not happen. The Bank of England and the US Federal Reserve printed extra cash lest there be a surge in demand in the last quarter in 1999. The Federal Reserve also created a special programme to support small businesses in need of loans to make Y2K-related repairs. These preparations proved unnecessary. In general the demand for cash was typical of other years and few loans were issued under the special programme (Schlesinger and McKinnon, 2000; and *Wall Street Journal*, 1999). Similarly, investment houses tried to anticipate Y2K-related activity in mid-1999 and then became cautious in late 1999 lest there be Y2K disasters (Tan, 1999). Again, both acts proved unnecessary as Y2K had only a modest impact on markets. Pharmaceutical suppliers worried about people stockpiling drugs and thereby creating a shortage, but other than a slight increase, stockpiling did not occur (Lagnado *et al.*, 1999). There was a slight decrease in demand for airline tickets on New Year's Eve (Gomes, 1999) but travel is relatively light on that particular night every year. The slight decrease in 1999/2000 can just as easily be attributed to people wanting to spend the 'Millennium Night' at home with their friends and families. Indeed, people participated in millennium celebrations in great numbers in US and in UK cities alike, seemingly without fear.

With reference to Downs's Issue Attention Cycle (1972), one might conclude that people simply became bored of Y2K. It was an obscure problem after all, which few people really understood in detail. Moreover, the story had been covered very thoroughly for well over a year by mid-1999, with hypothetical 'bang', yes, but little real or practical 'bang' to it, as few real problems ever materialized.

Relatedly, people might have considered various contingency plans but ultimately decided not to act on them because they were too expensive and inconvenient. If one reconsiders Figure 5.7, for instance, which claims people had or would change their behaviour in the light of Y2K, the test cases are somewhat problematic indicators of the public mood. For instance, it was relatively easy for one to say that one was *not* going to fly on a plane over the New Year period, particularly if one did not have specific plans to fly. As far as stockpiling food and water, that could mean anything from stocking a year's supply of spam in the back-yard bunker to simply picking-up a few extra groceries to ensure there are no disruptions to new year's parties. The only indicator that really required (unambiguously) changed behaviour was withdrawing all money from the bank. This is the only indicator that dropped over time. While the finance sector did publicize quite aggressively that it had fixed all Y2K problems, which likely contributed to the decreased anxiety levels in 1999, the 6 per cent of people who thought there was going to be a problem would still have to weigh the options. If one did withdraw all the money from one's account, where would one put the cash? Would it be any safer in the new location? After all, the central governments insure deposits up to a maximum in registered banks.

## The opinion-responsive hypothesis

The Opinion-Responsive Hypothesis (ORH) suggests that public attitudes shape regulatory regime content. In other words, that risk regulation is the way it is because that is how those affected by the risks want it to be (2001, 90).

Hood *et al.* discovered varying degrees of reaction to public opinion by government in the risk domains they examined. Figure 5.8 charts Hood *et al.*'s organizational framework for the hypothesis, which considers *active* through *passive* efforts to learn public opinion against strategies that are *aligned* with public opinion through those that are *not aligned*. Of the nine domains examined, Hood *et al.* discovered examples of each of the types, with the exception of 'perverse unresponsive'.

Equally pertinent to the discussion in this chapter, Hood *et al.* also note examples of strategies that regulators deploy when there is

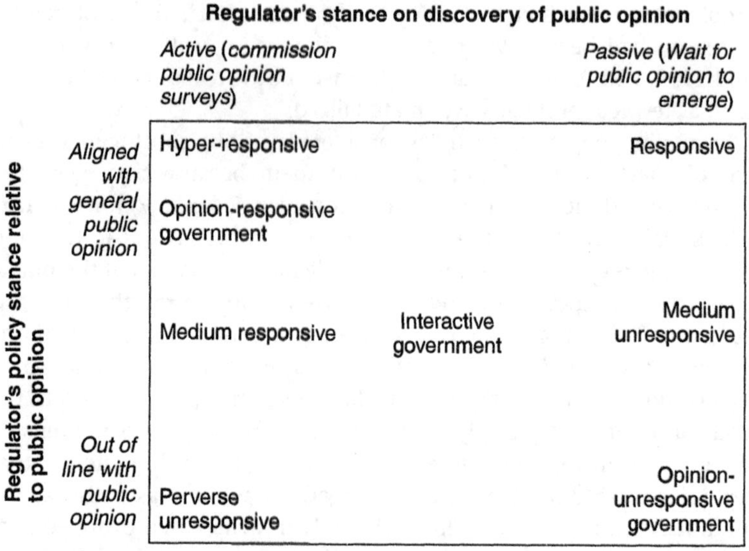

**Regulator's stance on discovery of public opinion**

|  |  | Active (*commission public opinion surveys*) | | Passive (*Wait for public opinion to emerge*) |
|---|---|---|---|---|
| **Regulator's policy stance relative to public opinion** | *Aligned with general public opinion* | Hyper-responsive | | Responsive |
|  |  | Opinion-responsive government | | |
|  |  | Medium responsive | Interactive government | Medium unresponsive |
|  | *Out of line with public opinion* | Perverse unresponsive | | Opinion-unresponsive government |

*Figure 5.8*   Observed regime content and opinion responsiveness, amended
*Source*: Hood *et al.*, 2001, 104. Reproduced with permission.

a gap between public opinion and these strategies. They specify four observed tendencies: disregarding the signals emanating from public opinion; paying 'selective' attention to public opinion; opinion-shaping strategies, which include education initiatives and 'spin'; and an attempt to balance government preferences with public preferences. They also note hybrids of all four strategies (103–9).

## To what extent can the governments' responses be explained by public opinion?

Neither the US nor the UK governments' reaction to Y2K can be explained by viewing government actions and public opinion in a strict causal relationship. The two agents experience a much more fluid and dynamic two-way interaction. Though the governments did not poll public opinion very often, they were greatly concerned by it. At the same time, for the most part, their actions suggest they tried to shape public opinion rather than follow it.[13] And while at times they aligned government operations with public reaction, particularly in 1998, at times they quite deliberately deviated from public opinion, ironically, in an effort to ensure a successful outcome overall. In short, once the governments put their Y2K operations in place, the size and structure remained by and large unchanged throughout the process, as did the zealous data-collecting style within government. The style of communications, however, was frequently closely aligned with public opinion, with an eye either to raising public anxiety, as in the 1997 and early 1998 phase, or to reducing public anxiety, as in the 1999 period, particularly in the United States. It was an interactive and iterative process.

Y2K was low on the public radar in 1997. No organizations were tracking public opinion because presumably few people outside the IT industry and senior management were particularly concerned about an obscure computer programming issue that was initially understood to be an internal operational matter. Indeed, even within the IT community, there were varying degrees of awareness of Y2K, and among those who knew of the bug, most were unsure of its consequences outside of the individual systems for which they were responsible.

At this time, media coverage was sporadic and mostly alarming, though at times reassuring and sometimes even cynical. One interview subject from the media suggested that the FT (and papers like the FT) tries to influence government policy (INT 31). Indeed, this seems the more likely reason for the alarming coverage in late 1997 and early 1998, particularly in the United Kingdom where there was sustained, alarming coverage from the FT immediately preceding the UK government's creation of Action 2000 and the Y2K Cabinet Committee, MISC 4. Figure 5.9 charts significant government interventions in relation to the alarming tone of the headlines on a national level. In both countries, the significant executive interventions in 1997/1998 occurred when the percentage of alarming headlines was high and there was an upward momentum.

At first glance, Figure 5.9 might suggest governments intervened to quell the anxiety. But the governments' underlying message—particularly in 1997/1998—was almost always alarming, not reassuring, effectively raising public concern from a previous low. Indeed, both executives seemed to be mirroring the media's tone. President Clinton, for instance, noted, 'with millions of hours needed to rewrite billions of lines of code and hundreds of thousands of interdependent organizations, this is clearly one of the most complex management challenges in history' (1998). Prime Minster Blair said, 'this is one deadline that is non-negotiable. Normal processes will not meet it. But by treating this as an emergency, we can make Britain one of the world's best prepared countries' (Blair, 1998).

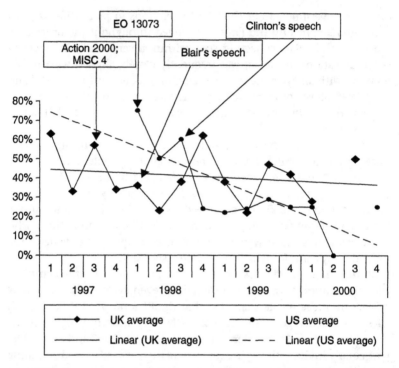

*Figure 5.9*   Tone of media coverage: percentage of articles with alarming head-lines sorted by country and by quarter, 1997–2000. (Significant Executive Interventions Noted)

The executives' interventions cannot be attributed solely to a reaction in the face of an alarmist press. They intervened significantly when *professional* opinions about Y2K more generally were at their most alarming—that is, among industry leaders, the respective legislatures and the media. Many players within these groups pressured the executives to act on Y2K, but there were differences. Arguably, for instance, the US executive was responding to a more aggressive Congress, which is generally much more effective at influencing policy and receiving media attention than backbenchers in the House of Commons. Indeed, the Congress was twice as likely as the House of Commons to be the source of a Y2K story in the media coverage examined for this research. The UK executive, on the other hand, seemed to be responding more directly to an alarmist press.

In many respects, the governments' approaches to Y2K can be viewed as a successful opinion-shaping campaign, particularly among SMEs, micro-businesses and the public. The governments could not police Y2K compliance across their entire infrastructures the way they did within government departments and agencies. They did not have the time, expertise, evidence, resources or legal authority to force the organizations in the national infrastructure to comply. Building on the momentum of a professional anxiety, both governments relied on the tools at their disposal in a complex network of loosely and tightly coupled interdependencies (Perrow, 1999, 4–6). It could convince, persuade, pressure and 'incentivize' organizations to act on Y2K, depending on the various organizational and cultural constraints from which they were acting.

Communications was critical to this intervention. The governments raised awareness—and in so doing raised anxiety levels—so that people would check and fix their systems and the government would thereby help to maintain stability in the face of the uncertainty. Being 'Y2K-compliant' became a label of good corporate citizenship. It represented taking one's business seriously; anything less than declaring 'full compliance' in most sectors of the national infrastructure was not acceptable. Paradoxically by 1999, after raising awareness and anxiety for over a year, the US government, in particular, was worried that it had done too good a job such that public awareness would result in hoarding and stockpiling (INT 54). Public enemy 'number one' was no longer the bug but the public itself, and they therefore embarked on a strategy of reducing public anxiety, with a good deal of success. Most executive interventions from mid-1999 onwards—Community Conversations, White House Roundtables, President's Council Summary of Assessment Reports—were decidedly less anxious in tone. While the UK Government was less concerned about public overreaction, they too issued reassuring messages towards the end of 1999, exemplified by the NAO's final report (1999b) and that of Action 2000 (1999).

It is impossible to state definitely exactly what impact government interventions actually had on public opinion but one thing was for sure, at a certain point changing *style* was all they did. While anxiety levels diminished throughout 1999, the governments were locked into large-scale bureaucratic responses trying to fix a peculiar, yet pervasive risk in a risk-averse environment. The governments' approaches were as thorough and expansive as the perceived problem was in 1998, but they were expensive and inflexible once they got started. To a degree, the nature of the technology and the typical framework for IT risk management within IT departments made such a response inevitable, particularly if one wanted to be certain that there would be no Y2K-related failures. As New Year's Day 2000 approached their extensive operations had deviated from the public perception of the risk. But in the light of the potential consequences of public panic—a reaction that seemed possible one year earlier—this deviation seemed acceptable. One might therefore describe both governments' reactions, particularly at the end of 1999, as *Perverse Unresponsive*, ironically, the only category Hood *et al.* failed to discover in their research.

## Some concluding thoughts on the opinion-responsive hypothesis

With respect to Hood *et al.*'s use of the media as a barometer for public opinion or as an insight into the 'flavour of public debate' it was inadequate for this research, particularly if we tie it to Gaskell's notion that increased coverage means a more negative opinion. By examining tone and content as well as volume, one gains a greater appreciation of the tone of public debate that volume of coverage alone could not demonstrate. In 1999, volume of Y2K coverage increased but public anxiety decreased. For the United States, the *tone* of the headlines seemed to be a better barometer for public opinion. In the United Kingdom, neither volume nor tone seemed to be an accurate barometer for public opinion. As noted, the FT in particular seemed to take a rather alarmist stance quite early on with respect to Y2K and never seemed to deviate from it. So when Y2K seemed to be less of a threat, the FT shifted focus to other countries that seemed either to be at risk or for which information on Y2K was sufficiently ambiguous that it could make ominous claims. In this case, volume, tone and content all have to be taken into account in order to understand the flavour of the debate.

In any event it is unclear how precise measures of public opinion of any one moment ever were. On the one hand, these polling numbers may confirm the observations in Chapter 2 about risk and the media—that people deny personal vulnerability and can be unrealistically optimistic. On the other hand, National Science Foundation's method of data collection, for instance, potentially undermined its own results. It is likely that the very act of asking people about certain risks triggers a heightened risk response. That is, there is a difference between asking someone, 'What are your hopes and fears for the upcoming year?' and 'Which of these services do you think will fail as a result of Y2K?' The latter would almost certainly reveal higher levels of anxiety about Y2K in particular but would not necessarily give an accurate insight into what people felt and how they would react in the light of those feelings. It is difficult to believe that by December 1999, 22–34 per cent of Americans actually thought that the services in Figure 5.5 would fail (that is, banks;

aviation; food and water; emergency services) and 8 per cent of UK citizens considered Y2K a 'serious threat' yet few modified their behaviour. In short, these polling numbers may be somewhat exaggerated and must be treated and interpreted with caution. The National Science Foundation's method cannot be attributed to Hood *et al.*'s framework. Rather, the problem highlights the difficulty in obtaining reliable information for the framework.

Moreover, government interventions can change public opinion and therefore public opinion is potentially never static. For example, when people failed to modify their behaviour at the end of 1999, was that because (i) government interventions convinced people that Y2K was under control and they need not modify their behaviour? (ii) of public inertia or ambivalence? (iii) of some other reason altogether? Understanding people's motivations is important for public policy debates. If the first interpretation is correct, then the government should indeed be congratulated for a highly successful public awareness campaign. If, on the other hand, the second interpretation (inertia or ambivalence) is correct, then what we saw in Y2K is troubling for future risk management initiatives. That is, in the face of potential operational shut-downs in which even some modest planning could save considerable inconvenience, the public tends not to deviate from its set pattern and routines (see Note 1). They become 'fatalists' in the Cultural Theory sense (see, for example, Hood, 1998)—'what ever will be will be'. These two very different interpretations of events run the spectrum of possibilities—one sees Y2K public relations as a success that should be repeated if necessary whereas the other one sees it as dismal failure, and which should be greatly modified should something like Y2K happen again. This research is inconclusive but exemplifies how difficult it is to draw conclusions about public opinion and/or the relationship between the variables.

By the same token, an opinion-responsive approach is a compelling one in Y2K because the government always had one eye on public opinion. The build-up of communications capacity on Y2K is instructive. Both the Cabinet Office and the Executive Office of the President had Y2K communications operations. Koskinen noted he deliberately kept new structures and staff to a minimum, yet chose to appoint a director of communications on the issue. Similarly, the Cabinet Office indicated it wanted to wrest control from an anxious media (Cm 4703, 2000). These communications departments created routines that monitored headlines and polls. They were not oblivious to what people were thinking about the issue. But at the same time, communications departments might be more usefully thought of as a forum for a 'conversation' with various sources (for example, media, Congress, an *imagined* public, key stakeholders) rather than as a monologue in which the government communicates information and the public readily accepts it. As the Social Amplification Framework suggests, views about risk emerge in a complex dynamic and interactive process that involves numerous sources (Kasperson, 1992). In this light, the strict division between government view and public view is perhaps too limiting.

In some respects, Y2K in particular complicates the question of government reaction in the face of public opinion polls even further because public opinion was not simply a way of legitimizing government action in a democracy; public opinion was also an operational variable, if you will. How the public felt and responded to Y2K had to be, at the very least, monitored. Practically speaking, it also had to be *managed* like any other operational issue. In this sense, communications need not be thought of as an add-on section to a policy initiative but rather as a policy tool in and of itself. If one 'scared' the small business owners into fixing their systems through unsettling messages, then the government would be helping to ensure stability among supply chains, for instance.

Like many other high profile events, Y2K was a difficult communications issue to manage. The governments were trying not simply to understand what opinion was *at the present* but also to anticipate public opinion *at various points in the future*, and for different though overlapping reasons. One reason was to avoid public anger should there be any breakdowns in the infrastructure. The other was to manage the problem of public reaction *causing* a major breakdown in the infrastructure. In both cases, the government could argue it still had public opinion in mind because it was trying to achieve the ultimate, long-term goal shared by the vast majority—stability in the infrastructure over the millennium. In this respect, the government might argue it was rising above the day-to-day politicking in favour of the more important longer-term goals.

If public opinion plays an important role in achieving operational ends in a high profile, 'live and active' issue, it is difficult to imagine that government reaction could ever be anything but 'interactive' as it tries to gauge, manage and respond to public opinion. The question it raises, however, is how useful is the framework if one sees manifestations of all eight strategies and ultimately settles for the middle one (that is, a bit of everything)?

Finally, it is also worth noting that in some cases it did not matter what the government did. It could never align itself with some elements of public opinion because some people simply did not trust the government. Peter de Jager's Y2K website, for instance, received over 500,000 hits in mid-1998 when Y2K was at its peak (Romano, 1998). Gary North was the author of a much used 'alternative' Y2K website. North was a Y2K pessimist who saw disaster and conspiracy throughout the run up to the year 2000. There were the (rare) cases of people who packed up and moved to the hills in fear of Y2K disasters (Wheelwright, 1998). There were also journalists that doubted the governments' line on Y2K. In an interview with an IT journalist, he noted that there were questions among the White House press corps concerning the trustworthiness of John Koskinen. Some felt that Koskinen had been appointed to reassure the public that everything would be alright but in fact even Koskinen was not fully briefed on the entire situation. In these views of Y2K, people believed that the CIA knew of dangers yet it would not discuss them with anyone (INT 52). Recall, for instance, 3 per cent of articles in the United States referred to potential acts of terrorism. In short, as noted in Chapter 2, some people's distrust of institutions—like government or large industry—or technology could not be overcome. Koskinen speculated this number was potentially as high as 20 per cent of the American population (INT 54). Trusted or not, however, powerful institutions did play a critical role in shaping government reaction to Y2K, a point to which we turn in the next chapter.

# 6
# The Interest Group Hypothesis (The Issue Network)

Each government had its own method of engaging with industry. The US government chose to tap into industry via trade associations and thereby secured a broad overview of the infrastructure. The UK government, in contrast, engaged directly with a select group of industry leaders and standardized (at a high level) the approach among them. Its effort to map sectoral interdependencies was ambitious, but flawed. In both countries, ultimately, larger players influenced the manner in which Y2K would be managed within their own sectors but they also had a responsibility to ensure continuity of service for their respective sectors and supply chains.

Parts of the Y2K story can be framed and understood in conventional interest group terms in which the US government is viewed through the pluralist lens and the UK government is viewed through the corporatist lens (Vogel, 1986). While such theoretical lenses are helpful, however, they are insufficient. There are key parts of the story that are not captured in such an interpretation, including the information-coordinating role governments played between sectors and the emergence of the loosely grouped IT Network that was not confined to an organization or a sector but perhaps more than anyone did influence the method and tools that were used to mange Y2K.

This chapter is organized according to the four dimensions of the Marsh and Rhodes Policy Networks framework: Membership, Integration, Resources and Power (Marsh and Rhodes, 1992, 251). It concludes by considering the extent to which conventional interest group theory helps to explain government/industry dynamics in the run-up to Y2K and the extent to which Issue Networks can further illuminate our understanding of this complex dynamic.

## Marsh and Rhodes's typology

Hood *et al.* (2001) use Wilson's (1980) typology of organized business interests to explore the Interests Hypothesis. I use Marsh and Rhodes's (1992) Policy Networks typology to describe and analyse group dynamics in the run-up to Y2K and will use the pluralist and corporatist lenses (Vogel, 1986) together with the Rhodes and Marsh's Issue Network to try to explain the interactions between government and industry in the United States and United Kingdom, respectively.

A policy networks approach is not without its critics. Dowding (1995) notes the overlap between each of the four organizing dimensions and criticizes how its advocates tend to explore the theoretical over the empirical. Similarly, Page (2001) notes the literature can be vague on how specific policy networks manage issues: the beginning and end of a Network is often difficult to specify.

The reason I elected to substitute Wilson's Interest Group approach with Rhodes and Marsh's Networks approach is the complexity of the Y2K issue. Wilson's typology assumes certain stability over time in the dispersal of costs and benefits that, at least at the outset, one cannot assume of the Y2K case. Y2K was relatively short-lived and dynamic. While, yes, there was considerable interaction within sectors, Y2K also brought organizations together that would normally not interact directly. The situation combined two pressures: the usual, sectoral-level dynamics with the unusual inter-sectoral dynamics. Large organizations that enjoyed privileged status within their own sector had a relatively diminished status in an economy-wide problem. Existing power structures were challenged. In short, everyone had a stake in ensuring a degree of Y2K compliance across the infrastructure and therefore no one stood fully outside of Y2K-related scrutiny. But nor could anyone fully police anyone else, particularly outside of one's own sector. Unlike Wilson's typology, Marsh and Rhodes's Issue Network (1992, 251) accommodates this complexity. Indeed, such an approach has a precedent in the networks literature: Read (1992), for instance, describes two network types simultaneously providing context for an issue (smoking)—the inner groups and the broader issue network.

This chapter focuses primarily on the loosely integrated Issue Network that emerged from across the infrastructure as a response to Y2K. It went beyond the lines of conventional sectors. That noted, there are three other networks that will also be referred to at different points of the chapter that provide further context: the (relatively more) tightly integrated networks (i) within sectors and (ii) within departments and agencies; and finally, (iii) the more loosely integrated network within the IT community. By exploring multiple networks the chapter provides a greater portrayal of the complex context that helped to shape the respective governments' reactions.

## Membership

Membership of Issue Networks is large and encompasses a range of affected interests (Marsh and Rhodes, 1992, 251). In many respects the Y2K issue touched on all sectors, public and private. As the NAO noted, 'there are very few areas of modern life that are not touched by information technology. The millennium threat [was] a business-wide problem that affect[ed] everyone' (NAO, November 1999). That noted, decisions were made about which sectors would be identified for particular attention when the governments ring-fenced the sectors in the UK/NIF and the US/WG. Table 3.2 (here, Table 6.1, Action 2000, 1999) from Chapter 3 is recalled here and highlights those sectors that were included in the Y2K Issue Network.

On the whole, Table 6.1 shows that labelling is slightly different but the entries are similar. In some cases US/WGs subsumed many of the functions that the UK/NIF sectors treated separately. For example, the US/WG Transportation included five UK/NIF categories—Rail Transport; Air Transport; Road Transport (Local Government); Sea Transport; and Bus Transport.

While the lists in Table 6.1 are similar they are by no means identical. Some entries reflect arrangements that existed in one country but not in the other, such as the United States's inclusion of state government.[1] In other cases, some functions were simply omitted. For instance, the US Government included Defence and International Relations among the Working Groups, whereas the UK government decided not to include the Ministry of Defence nor the Foreign Office in the UK/NIF explicitly, despite both having sizeable Y2K operations in place. (The MOD, for example, accounted for 39 per cent of the UK Government's overall Y2K expenditures (Cm 4703, 2000, 76)).

The omission of the IT sector and SMEs in the UK/NIF is particularly noteworthy. While the UK/NIF might have included the Federation of Small Business, for example, there was obviously no practical way to show that SMEs were going to be 'business as usual'. Therefore SMEs were embedded within the overall approach of the UK/NIF—each member of the UK/NIF would have to ensure its own supply chain, which would necessarily include many SMEs. In some ways this approach was

*Table 6.1*   Critical sectors represented at the National Infrastructure Forum (UK) and the Working Groups (US) (repeated)

| Tranche | National Infrastructure Forum – UK (25) | Working Groups – US (26) |
|---|---|---|
| 1 | Electricity | Benefits payments |
| | Gas | Building and housing |
| | Fuel supplies | Consumer affairs |
| | Telecommunications | Defence and international security |
| | Water and sewerage | Education |
| | Financial services | Emergency services |
| 2 | Essential Food and Groceries | Employment-related protections |
| | Rail transport | Energy (electric power) |
| | Air transport | Energy (oil and gas) |
| | Road transport (Local government) | Financial services |
| | Sea transport | Food supply |
| | Hospitals and health care | Health care |
| | Fire service | Human services |
| | Police | Information technology |
| | Broadcasting | International relations |
| | Local government | International trade |
| 3 | Sea rescue | Non-profit organizations and civic preparedness |
| | Weather forecasting | Police and public safety |
| | Post and parcels | Small business |
| | Welfare payments | State and local government |
| | Flood defence | Telecommunications |
| | Criminal justice | Transportation |
| | Tax collection | Tribal government |
| | Bus transport | Waste management |
| | Newspapers | Water utilities |
| | | Workforce issues |

*Note*: The UK/NIF sectors are divided by tranche. The US/WG sectors are listed in alphabetical order.
*Source*: Action 2000, 1999; President's Council on Year 2000 Conversion, 2000.

paternalistic; it encouraged large organizations to pressure the SMEs in their supply chains. The United Kingdom's omission of the IT sector is also ironic. Many argued that one reason that Y2K had been neglected in the early and mid-1990s was because IT did not have a sufficiently high profile at the executive level in most organizations. Despite this problematic legacy, the UK government still chose to see IT as a function that supported organizations rather than as a sufficiently critical sector in its own right, such as gas or electricity.

As noted in Chapter 3, with respect to government departments and agencies, both countries ultimately included all within the Y2K programme, as well as including critical and non-critical systems. But it was not always that way. Initially the US government targeted 24 departments (and certain agencies) deemed critical but only expanded to include smaller departments and agencies when Congress and the GAO demanded better information on government progress. The UK government always included all departments but only added non-critical systems to the strategy later in the process.

## Integration

Integration refers to the overall coherence of the network on the Y2K issue. As measures of integration I consider (i) the degree and stability of agreement among members of the network on Y2K management strategies; and (ii) the degree of interconnectedness of IT systems, their users and their systems providers in the execution of their Y2K strategies. In doing so I examine both horizontal and vertical integration, where horizontal refers to measures across representatives of the infrastructure and vertical to measures between the operational front lines and the executives.

### Horizontal Integration

Representatives of the infrastructure in the United Kingdom were brought together earlier than those in the United States. The House of Commons Science and Technology Committee hearings on Y2K (held between autumn 1997 and spring 1998) were an effective read of opinions about Y2K across the infrastructure, with the witnesses providing testimony from all three sectors (public, private and not-for-profit). This comprehensive overview eluded the US Congress until the Senate established the special committee on the Year 2000 in April 1998. The most active committees were the Subcommittee on Technology and the Subcommittee on Government Management, Information and Technology, which held joint hearings on the issue. But even then their witness lists were largely confined to IT consultants and staff from government departments. The Congress had been active on the question of Y2K two years before the establishment of the Y2K committee, but congressional activities lacked coherence. The approach to Y2K within the Congress was largely on a committee-by-committee basis.[2]

Moreover, the witnesses that appeared before Congress were frequently trade representatives. Their testimonies were frequently broad-brush generalizations; they lacked specifics and sometimes the credibility that a CEO from a major corporation would have. Tables 6.2 and 6.3 summarize the witness lists for the United Kingdom's Science

*Table 6.2*   Y2K testimonies before the UK Science and Technology Committee, November 1997–March 1998

|  | Live and written testimony | |
| --- | --- | --- |
|  | Direct* | Indirect** |
| Critical sectors | 17 | 8 |
| Government | 7 |  |
| Regulators | 10 |  |
| SMEs |  | 1 |
| Societies** (all indirect) |  | 5 |
| Legal | 2 |  |
| Property management | 1 | 1 |
| IT, including consultants | 15 | 2 |
| **Total** | **52** | **17** |

*Submitted by a company/organization from within the category identified.
**Submitted by a buffer organization, for example, an interest group; a trade association.

*Table 6.3*   Y2K testimonies before Congressional Subcommittee on Technology and the Subcommittee on Government Management, Information and Technology, March 1997–March 1998

|  | Live testimony | |
| --- | --- | --- |
|  | Direct* | Indirect** |
| Government*** | 14 |  |
| Legal | 1 |  |
| IT, including consultants | 7 | 2 |
| **Total** | **22** | **2** |

*Submitted by a company from within the category identified.
**Submitted by a representative organization, for example, an interest group or a trade association.
***Includes EPA and FAA, which are regulators but are counted as government because they were asked largely to testify about their own systems not the systems of those they regulate.

and Technology Committee and the United States's Subcommittee on Technology and the Subcommittee on Government Management, Information and Technology.

While they may have come together earlier, members of the network in the United Kingdom did not necessarily agree on strategy except to the extent that many wanted at least for government to do something. Participants submitting memoranda were given specific questions to answer. Considering the responses of those who answered the second of the questions, *do you think that the government has done enough to raise awareness of the problems associated with the date change or to encourage action to avert problems? What more should be done?* I obtained the results in Table 6.4.

Even within the context of a fairly scripted question, the results show different opinions about what the government should be doing. In general terms, some recommended that the government simply keep the public and businesses informed (gathering and disseminating information) while others recommended stronger intervention (modify behaviour). Specifically, the most common themes among the respondents were a call for government to coordinate information-sharing across sectors (10), raise awareness (7), ensure critical services (7) and help SMEs in particular (6). There were fewer calls for legislation (4), tax incentives or additional funding (3) or Y2K training incentives or opportunities (1). The US testimonies are less structured and they offer less explicit advice, which makes degree of agreement more difficult to analyse. The witnesses from the IT industry chose to describe the consequences of potential Y2K-related problems through the implicit, if not explicit, declaration by the witnesses that the problem was serious, and that behaviour modification across all sectors was necessary to avoid serious problems.[3] While there was no clear sense of agreement about what government should do, many agreed that uncertainty was running high. Members of Congress from both parties concluded during this period that the Executive should play a stronger role, both within government and outwith.

*Table 6.4* What the participants at the Science and Technology Committee hearings wanted the UK government to do about Y2K

|  | Gather and disseminate information | Set standards | Modify the behaviour of government or others |
| --- | --- | --- | --- |
| Total number | 56 | 28 | 40 |
| Expressed as ratio | 4 | 2 | 3 |

The FAA, for instance, was slow to become actively involved in ensuring the integration of the aviation sector. Somewhat ironically, when the FAA started to organize 'Industry Days' it discovered that there was a high degree of agreement on strategy already; in fact most organizations were making similar levels of progress. During interviews, staff noted that many organizations in the sector had been active on the Y2K problem for some time[4] but there had been little information-gathering at the meso- or macro-level, so no one really knew what the status of the sector as a whole was (INT 38). The FAA also facilitated contact between sectors. Industry Days included presentations from representative from a cross-section of major sectors. More than anything, this reassured participants.

In fact, the FAA itself was (and is) a highly decentralized operation. The FAA had seven Lines of Business (LOBs), which staff described as largely independent from each other, given the distinctive nature of their businesses (for example, Commercial Space; Airports; Security; Regulation and Certification; Air Traffic; Research and Acquisitions; Administration). (See, for example, INT 29; INT 45.) Moreover, not only was the FAA organizationally decentralized, it was also geographically decentralized. Regional offices and the air traffic control centres existed all over the country. One interview subject noted that many of the GAO's early criticisms were in fact misleading. She noted that while there may not have been a central IT inventory, for instance, she was confident that each LOB had one. Similarly, she noted the GAO may criticize the FAA for failing to have a joined-up purchasing strategy but in fact the technology requirements of each LOB are dramatically different and therefore there would be few efficiency gains by attempting such coordination (INT 45). It might be noted, therefore, that while there may have been little integration between LOBs, integration seemed unnecessary prior to the Y2K problem.

Integration within departments and agencies was adversely affected by the outsourcing of IT service providers. While IT may have been becoming more integrated into government work routines, those who were providing IT support in one sense were becoming less integrated. In the early 1990s, during John Major's Market Testing programme (Cm 1730, 1991) IT services had frequently been outsourced. This trend continued with PFI and PPP initiatives. Dunleavy *et al.* note that IT departments in government were reduced to small advisory units, while contracting out IT rose from 23 per cent of all civil service IT budgets in 1993 to 30 per cent in 1995 to 54 per cent in 2000 (2001, 9). Contracting out had long been a strategy in the US government. Margetts notes that by 1994 the US federal government was using private sector personnel in a huge variety of contracting arrangements, ranging from the hiring of individual contractors to the contracting out of departments' whole operations. Nearly 50 per cent of the US federal government's information technology work was accounted for by commercial services. The cost of information technology staff as a percentage of information technology expenditure dropped from 41 to 22 per cent over ten years (Margetts, 1999, 21). This fragmentation had operational impacts.

Ultimately, the fragmentation between areas proved to be a challenge for testing when, after working largely in isolation, programmes had to come together and conduct end-to-end testing, which necessarily involved a degree of participation and cooperation across programmes and departmental lines to which staff were not accustomed. But this outsourcing had other consequences. Among departments that had outsourced IT, there was not necessarily an opinion on Y2K. Therefore it was not really a case of detecting agreement between service provider and IT user. By outsourcing their IT services (and arguably much of their IT intelligence) many departments were unable to form an opinion on Y2K and had to accept whatever Y2K advice they received. This was not strictly the case. Clearly the BLS IT staff, for instance, had a strong view about how to manage Y2K. But many other smaller agencies[5] whose staff I interviewed seemed less involved in the Y2K process. Directives came from the lead departments and agencies simply passed them on to their IT service providers.

That noted, Margetts's and Dunleavy's observations about greater private sector involvement do not necessarily lead to a conclusion that IT staff were as disconnected from departments and agencies as Dunleavy and Margetts might suggest. As noted in Chapter 3, while it is clear that more IT staff were on the payroll of private companies, it is also clear that many staff continued (and continue) to work in government departments and agencies alongside public servants with sometimes little distinction between private sector staff and public sector staff.

**Vertical integration**

There was less agreement between the UK government and SMEs. Both governments assumed a tentative agreement with leaders of industry about how to proceed on Y2K. By and large, the UK government's response was similar to that recommended by representatives from the critical sectors. The government applied considerable pressure to its own departments to comply and created the UK/NIF to facilitate information-sharing and apply (indirectly) pressure on organizations to move towards compliance through sector-level, self-regulating groups with a degree of transparency. Despite SMEs seeming the most vulnerable to Y2K-related problems, the requests of the Federation of Small Businesses were largely ignored, notably, a delay on the single currency, tax incentives and legislation regarding embedded systems. In the United States, the government believed there were sufficient market incentives in place for industry to move towards compliance and stressed this point more often in publications and interviews. The government was also content to allow industry associations to assume the lead on Y2K and thereby not intervene too directly in sector-level dynamics. Industry did not challenge this position. That noted, leaders from certain industries did request protection from lawsuits should they offer Y2K assurances in good faith but turn out to be wrong (INT 54); this request was accommodated.

The US government's approach to the infrastructure was less interventionist and therefore the degree of stability across the infrastructure was somewhat less certain. Information in the United States was drawn from large numbers of industry surveys (self-assessments) sponsored, collated and filtered by industry associations. (See Table 6.5 below.) Not all organizations were compliant but most were. The UK government's approach seemed 'selectively' interventionist, in that it targeted

*Table 6.5*  Methods for determining Y2K Compliance across selected sectors of the US infrastructure

| Sector | No. of organizations participating/targeted for Y2K-compliance initiatives | Method | Lead/coordinator |
|---|---|---|---|
| Electric Power | 3000 electric power companies, representing nearly 100% of the industry | Surveys | NERC (leading trade association) |
| Oil and Gas | 1250 organizations, representing 93% of oil and gas consumed | Surveys | Federal agencies/ industry Groups |
| Water | 4000 organizations, out of a possible 190,000 | Surveys | Three leading Water Associations |
| Essential Food and Groceries | 55% of independent operators | Surveys | National Grocers Association |
| Hospitals and Health Care: Pharmaceuticals | 1500 organizations (manufacturers, distributors, packagers) | Surveys | Food and Drug Administration |

*Source*: President's Council on Year 2000 Conversion: August 1999 Report.

only the top companies; but it was a stable, agreed and interconnected arrangement. As noted in Chapter 3, the UK/NIF approach was more coherent but less transparent. The NAO and Action 2000 made generalized claims about the sectors, even when for reasons of complexity and volume such claims were contestable[6] (Table 6.6).

*Table 6.6* Number of organizations participating in UK/NIF by selected sectors

| Tranche | National Infrastructure Forum sector | No. of organizations participating in UK/ NIF-sponsored in-depth audits | Comments |
|---|---|---|---|
| 1 | Electricity* | Generation: 5 Transmission: 1 Distribution: 12 Supply: Electricity sector as a whole | Described as 'the major players' |
| | Gas** | 3 critical companies; 44 non critical (shippers and suppliers) | Included 'the majority' of companies in transportation and supply |
| | Telecommunications | 19 | Major players |
| | Water and Sewerage* | 26 | All water and sewerage companies |
| | Financial Services | 250 med impact; 150 as high impact; | *c.* 7500 institutions in total in the sector; remainder considered low or no risk |
| 2 | Essential Food and Groceries | 36 | Top 12 retailers represent 85% of industry; but smaller coops in remote areas were also assessed |
| | Rail Transport | 30 | |
| | Air Transport | 15, then 14 | Represented 80% of UK industry in terms of passengers or cargo carried One company dropped out during the process |
| | Hospitals and Health Care (Pharmaceuticals) | 10% of the industry | Normal inspection processes (every two to four years) included Y2K questions |
| | Fire Service* | 50 | All 50 in England and Wales |

| | Sea Transport (Ports) | 31 ports | 15 largest ports plus further 12 from top 50 were inspected; further 4 small lifeline ports |
| --- | --- | --- | --- |
| 3 | Newspapers | 9 | 5 national and 4 regional |

*This example is taken from the England and Wales entry only.
**This example does not include Northern Ireland entry.
*Source:* Action 2000, 1999.

While there were varying degrees of horizontal integration among IT staff and the civil service, it is even more doubtful how much vertical integration existed in this area. As noted, those responsible for IT were rarely on the executive. The US government tried to increase the stature of IT when, by way of the Clinger-Cohen Act, it mandated each government department to have its own CIO that would report directly to the head of the department. It also mandated the creation of the CIO Council. But these were new initiatives of the mid-1990s and while they helped the status of IT staff, they serve to illustrate their lack of status before that time. In some cases this lack of recognition continued throughout the Y2K period. For instance, the FAA refused to appoint a CIO that reported directly to the head of the FAA, despite continual criticism from the GAO on the matter.[7] Staff noted when at first the FAA tried to have a CIO but it did not work; the LOBs performed too discrete a function (INT 29; INT 45). Similarly, the BLS IT staff seemed unable to penetrate the executive with its (counter-culture) Y2K advice. IT similarly lacked status in the UK Government. While Deputy Prime Minister Heseltine tried to strengthen the role of central IT offices in the mid-1990s, it was by way of reversing a trend of cutbacks in the 1980s and 1990s (Margetts, 1999). And while the ONS did not particularly offer contrary advice to the centre, it too seemed to be driven largely by the templates sent by Cabinet Office. If there was vertical integration within the civil services, it was top down only.

In sum, the 1997 period is characterized in both countries by a concern over Y2K but a lack of formal integration or agreement on how to proceed. Operations were highly decentralized and fragmented both within government departments and agencies and across the infrastructure, both horizontally and vertically. By 1998, the US government tapped into information *via* existing trade associations that represented critical sectors. The UK government identified key organizations and penetrated them to a greater extent. Ultimately the US government had a broad overview whereas the UK government had more of an in-depth view of a (powerful) few.

## Resource exchange

Resource exchange is central to Marsh and Rhodes's concept of Networks; it is the glue that keeps the members together. In this section I will examine the concept by considering the manner in which relationships were prioritized and understood as well as by providing anecdotal evidence about the strained manner that characterized resource exchange in the run up to Y2K.

### Conceptualizing resource exchange for the infrastructure

Action 2000 decided to divide the sectors by tranches (see Table 6.1 in the section on Membership). Action 2000 described the tranches as groupings based mainly on their dependencies (that is, resource exchange) on each other (Action, 2000, 1999, 4). The sectors with the highest level of dependency on them were placed in Tranche 1. Tranche 1 had to report its Y2K status first, with Tranche 2 to follow a few weeks later and so on (Action, 2000, 1999, 4). For example, an organization in Tranche 2 could not claim to be Y2K compliant until all organizations in Tranche 1 had already demonstrated that they were compliant.

The project was ambitious (and unique in the world, in fact) but arguably, the top-down, flow-chart approach between the tranches does not capture the two-way, interactive complexity of the relationships. One shortcoming in particular would resonate with critical theorists. *Welfare payments,* for instance, is part of tranche 3 yet it is unclear, one, why *welfare payments* was not grouped with *financial services* (whose membership was dominated by banks and assurances companies) and, two, why *financial services* was considered tranche 1, compared to *welfare payments* being ranked as tranche 3. The grouping and placement within tranches suggest sectors were grouped according to the membership (for example, banks) rather than their subject (for example, money) and also implies the movement of some people's money is more important than that of others, as one member of the Public Accounts Committee suggested.

Relatedly, the UK/NIF attempted to fill a gap that had emerged in many organizations. One interview subject from NATS noted, for instance, that despite its dependence on critical service providers in other sectors it has never acknowledged or managed them particularly well. As noted, NATS ended up developing an expensive contingency plan that overlapped considerably with the responsibilities of other sectors (INT 63).

*Table 6.7*   Departments that were members of the most US/WGs

| Department/agency | Number of US/WGs for which the department was a member (out of a total of 26 US/WGs) |
| --- | --- |
| Agriculture | 16 |
| Health and Human Services | 11 |
| Department of Commerce | 10 |
| Department of Transportation | 8 |
| Environmental Protection Agency | 7 |
| Treasury | 7 |

In the United States, the interdependencies were not mapped as such. One might instead consider the dependence between sectors by looking at multiple memberships in US/WGs. Table 6.7 lists the six departments that were members of the most US/WGs.

The multiple memberships that departments held suggest the complexity of resource dependency. But it also suggests the difficulty in capturing this complexity, especially when one attempts to coordinate activity *via* existing institutional arrangements. Some organizations, for instance, found themselves participating in groups in which they had little in common with other members other than the fact that they were associated with the same federal department. The Transportation Group had a very wide remit, for instance. Others found themselves repeating similar messages to a large number of US/WGs (for example, Agriculture participating in 16 groups). In some cases the lead agency for a particular group seems questionable (for example, Veteran's Affairs leading the Pharmaceutical Group as opposed to the Food and Drug Administration or the Department of Health and Human Services, suggesting the US/WG might reveal a bias towards pensioners' issues).

The scope of the UK/NIF also has limitations. The Ernst and Young report (the blueprint for the Action 2000/UK/NIF approach to Y2K) was commissioned by the Cabinet Office and ordered on a short and fixed deadline. The report concedes at the outset that representatives from only a small sample of organizations were interviewed in developing it (sometimes as little as one person per sector), that conclusions about process dependencies were based largely on the opinions of these individuals and that differing views emerged during the interview process (3). Moreover, Ernst and Young conducted the research when arguably anxiety was at its peak (4 May 1998, to 30 June 1998) (7),[8] an environment ripe for exaggerated claims.

Not surprisingly, then, when the Ernst and Young report was implemented the project was constrained by the small sample of references. The report lists critical functions for the national infrastructure, but it fails to regionalize the picture. For instance, the south of England occasionally experiences flooding, its homes frequently depend on gas lines and the train network is complex. Ernst and Young identify all these processes as 'key' processes. Yet, none of these problems exist (for all intents and purposes) in Northern Ireland. Nevertheless, the government pressed each region to follow this same checklist. Hence, Northern Ireland was forced to divert resources to investigate issues that were highly unlikely to materialize. Northern Ireland eventually distanced itself from the UK/NIF, established its own NI-NIF, and worked predominantly in that forum.

## Resource exchange within government: negotiation under stress

The outsourcing and privatization noted above added to the cost, time and complexity of Y2K compliance. Y2K compliance had not been foreseen nor negotiated in many of the original IT service contracts with externals. Indeed, most departments and agencies had to pay a supplement to the organizations that provided them with IT support to perform any Y2K-compliance related work. One UK agency's Y2K coordinator noted that his external service provider saw Y2K as a business opportunity. The coordinator felt he spent most of his time keeping the external contractor's enthusiasm, and subsequent Y2K billing, down (INT 19).

IT manufacturers' Y2K billing practices varied. Some provided Y2K compliance 'patches', downloadable free of charge from their websites. When work was more involved, however, IT manufacturers charged for Y2K compliance work. IT manufacturers argued that Y2K compliance was considered an enhancement, and therefore it had to be negotiated. IBM, for instance, testified to the Standing Committee on Science and Technology that 'upgrading' was common in the hardware/software business and Y2K was no different than other 'upgrade' situations (Standing Committee on Science and Technology, Chapter 4, first page). Indeed, some large organizations capitalized on the situation by creating 'software factories' that specialized in making clients' software Y2K compliant. IBM, Capgemini and Unysis had 23, 21 and 10 of such factories, respectively (Nairn, 1998).

In one case it was not clear who owned the system—the government or the external service provider. With the considerable demand for the external contractor's services, and with the market pushing the cost of IT labour up the department staff noted that the external IT service provider was prepared to wait until the UK government assumed responsibility (INT 4). In the end the UK government capitulated and paid for the 'upgrade'. One civil servant described the IT suppliers as 'opportunistic', taking advantage of the government's (and the department's) vulnerability (INT 9).

Some respondents claimed that IT service providers were deliberately evasive about the reliability of existing systems as a sales pitch for the new 'Y2K compliant' version of their product. Interview subjects noted that buyers could either accept the highly qualified guarantees about Y2K compliance of older products or else pay for (sometimes) expensive exploratory work; or they could simply agree to buy the latest version of the software. In an environment that was seeking certainty, the latter was often the preferred (and easiest) option.

In sum, there were considerable resource dependencies across the national infrastructures but in the main, they were poorly documented and managed and were difficult to conceptualize, especially given the complexity and the time constraints.

## Power

In this section I draw from Rhodes's concept of power-dependence to examine inter-organizational dependency, how the 'the rules of the game' were determined and how discretion was used in interpreting those rules. This section uses these concepts to examine three critical contextual relationships for this research. Namely, the relationships: (i) between the government and industry; (ii) relationships within government departments; and (iii) within the IT sector.

### Interactions between industry and government

UK/NIF audits represented a confluence of interests between the Government, the regulators and industry. It was unlikely that larger players in the UK/NIF required the additional government-sponsored audits to help them achieve Y2K compliance. Most had the resources to manage Y2K problems on their own. In fact, the CAA first learned of Y2K concerns from the industry in 1996 (INT 22). The major players within the aviation industry committed considerable time and effort to ensure Y2K compliance in advance of the new millennium. The Air Transport Association estimated that internationally airlines spent $2.3 billion on

Y2K. BA, for example, the largest service provider in the United Kingdom, spent £100 million and had 200 people check 3000 systems, 50 million lines of computer code, 40,000 PCs and printers and 800 applications (Bray, 1999). The larger players did not do this massive check at the insistence of the CAA but rather because they recognized the potential seriousness of the problem and were concerned about the volatility of their clients vis-à-vis the Y2K issue. This approach was quite consistent with the CAA's original position: each organization is responsible for its own compliance, like any other safety issue. By the time Number 10 and the Cabinet Office established the UK/NIF in late 1998 the major players in the industry had gathered sufficient information to conclude that the problem was under control, particularly with respect to health and safety systems (Bray, 1999).

However, the terms of the UK/NIF dictated that there had to be a negotiated settlement between government watchdog and industry. Recall that the CAA never did receive confirmation from 22 organizations concerning Y2K compliance. The CAA did not act on this deficit, though it had the right to, because the CAA felt the risk was negligible.[9] The UK/NIF, on the other hand, had zero tolerance for such overt risk-taking: all who participated had to follow the standard steps to demonstrate that it would be 'business as usual', including a mandatory third party audit of all UK/NIF-participating organizations. This would allow the government to declare aviation would be 'business as usual'. Hence, the CAA contracted AEA Technology to conduct audits of a small but reliable group of large players in the industry that represented 80 per cent of passengers. This initiative was a practical response to meet Number 10's and the Cabinet Office's request for a form of demonstrable compliance. Ironically, although the *Safety Programme* only audited between 8 and 10 per cent of organizations, the top 15 companies were certain to be among them. In any event, judging by outcome, it was the small to medium-sized enterprises (SMEs) with older, less sophisticated systems and with fewer resources that were having a harder time achieving and demonstrating Y2K compliance and whose behaviour required modification by way of a firm reminder from government regulators.

The FAA had similar problems, but found different solutions. During interviews, staff indicated that a senior Senator actively involved in the Y2K process initially insisted that any organization that failed to produce a Y2K compliance certificate would have to have their licences revoked. FAA staff resisted strongly. They argued it was impossible to contact 'every mom and pop shop' licensed to fly a plane in the United States. There were too many, several of which were remote and in any case seasonal (summer only) operations. Even if one could send them all a letter, one could not guarantee a reply, let alone an audit. In any event, without any concrete evidence that the company's systems would fail, one could not revoke the organization's licence. Eventually, the senior Senator and the congressional committee removed the request and accepted the FAA's proposal—that the FAA would send a letter to each organization with a license and would try to secure as many replies as possible (INT 45).

The Y2K Act simultaneously protected the IT industry from lawsuits and devolved authority in a complex environment to large, private organizations to help facilitate stability in each of the main sectors. The Y2K Act and the Information and Year 2000 Readiness Disclosure Act limited the aspirations of legal firms wishing to cash-in on Y2K lawsuits. The Association of Trial Lawyers of America (ATLA) was opposed to the Y2K legislation. In two separate editorials *The New York Times* made five separate points against the legislation, focused mainly on protecting people's right to sue and the fact that the legislation let the IT industry off too lightly (NYT Editorial desk 1999a,b). The dispute over the legislation was particularly noteworthy because it pitted two interest groups against one another that typically support the Democrats, ATLA and the Information Technology Association of America (ITAA). Yet ITAA was not alone in supporting the legislation (Simons, 1999a,b). Eighty large companies and trade organizations also supported the Y2K Act, including the National Association of Manufacturers and the US Chamber of Commerce (Simons, 1999a,b). The Y2K Act helped the IT industry avoid lawsuits.[10] More importantly, it helped large organizations secure the supply chain and help the stability of their own sectors. With a complex environment to harness, large organizations could provide advice to smaller operations to help ensure they were compliant without the large company exposing itself to a lawsuit should its advice be incorrect. Indeed, after the Year 2000 Information and Readiness Disclosure Act was passed into law, several large organizations, such as those in the telecommunications sector, started sharing information with smaller companies and suppliers. Similarly, leaders of the electric power industry began a series of regional conferences for local distribution companies in which they discussed problems, particularly with embedded chips, as well as tested protocols and contingency planning (President's Council on Year 2000 Conversion, 2000, 6).

Such work was not exactly philanthropic. The executives from these large organizations were protecting their own industry, which largely had to protect their reputation 'jointly', by sector. Those larger organizations pressured their supply chains to conduct audits. Some smaller organizations had no choice but to incur these costs even if they felt that they did not have any (or few) Y2K vulnerabilities. A number of US banks forced companies to undergo audits before they could receive loans while the Securities Exchange Commission (SEC) made all public companies disclose Y2K preparations (Taylor, 1998). In one conversation I had with an IT executive from an SME[11] he noted 'feeling forced' by his primary customer into agreeing to an expensive Y2K audit that cost $750 K.

That noted, some smaller organizations still retained sufficient power to challenge the rules. One medium-sized UK bank in the finance sector in the UK/NIF fell behind in its Y2K compliance. Other organizations in the group argued that it should be named, lest the reputation of the whole industry be damaged. The Financial Services Authority (FSA) was reluctant to name any organization voluntarily participating in the UK/NIF and in any case, when pushed, the failing organization argued that no one could prove negligence because nothing had yet failed (Committee on Public Accounts (PAC), 1999, 2).

## Government departments and agencies

The centre's efficacy at ensuring Y2K compliance was facilitated by the governments' declared strategies that government programmes would not be disrupted as a result of Y2K. The strategy tips the balance away from a public service-driven Y2K strategy, in which civil service managers determine the appropriate level of Y2K compliance ('as ready as it can be'), to a 'citizen-friendly' strategy, in which success is determined by continuity of service from the public service user's perspective. That noted, this pressure rarely came directly from citizens but rather from governments anticipating citizens' interests.

Yet in so doing the governments minimized operational flexibility. Critical as well as non-critical systems were ordered to be Y2K compliant. (See, for example, NAO, 1999b, 34.) The Cabinet Office- and OMB-devised reporting template required progress reports on *all* systems that were listed on the systems inventory, as well as status reports on the Y2K compliance of department and agency suppliers and the readiness of departmental and agency contingency plans. As a result, departments and agencies were directed towards an 'inventory, fix, test and audit everything' operational plan. It echoes the conventional IT approach to risk management noted in the literature review: identify; segment;

eliminate. Like the 'no material disruption' strategy, this approach reduced ambiguity from the centre's perspective.

However, as one senior civil servant in the United Kingdom noted, given the perceived high degree of uncertainty, departments and agencies were relieved to be given clear marching orders (INT 1). In the United Kingdom, the anxiety within the senior civil service went beyond local operational shutdowns. One senior civil servant noted that he and others were aware that the Modernisation Agenda formed part of the backdrop to the Y2K operation. There was a feeling that had there been any significant systems failures as a result of Y2K, they would have undermined the government's credibility in delivering the programme, not only because it would expose IT vulnerability but because it would reveal a weakness in the government's (in)ability to manage it. This resulted in some pressure at the departmental level (INT 1). Similarly in the United States, interview subjects noted Y2K was certainly seen as one of the first test-cases for the CIO Council, which was created just as the Y2K issue was gaining momentum.

The same thoroughness cannot be noted for the departments' and agencies' approach to their (non-IT) suppliers. Departments and agencies pursued critical suppliers vigilantly. In fact, most large organizations, like the government, had a Y2K plan in place and at a minimum were able to reply to correspondence. Nevertheless, often the letters the suppliers wrote were so heavily qualified that they were 'not worth the paper they were written on' (INT 11). Moreover, the response rate from small to medium-sized suppliers was much weaker.

Despite the constraints on the operational front line, there were occasions when government departments and agencies used the rules of the game to their own advantage. First, in the United States, where Y2K proposals were funded centrally, programme areas rushed proposals under the auspices that it was a Y2K request when in fact it was just a desire for new equipment.[12] This opportunism also occurred in the United Kingdom; however, the requests were not funded by the Treasury but by the departments themselves.[13] Second, many staff made personal gains. Most civil servants interviewed who worked directly on Y2K programmes 'rode the Y2K tide' and enjoyed many of the benefits. Many staff interviewed for this research were promoted, for instance, either to do Y2K work, during their Y2K work, or just as the Y2K work was completed. Similarly, many staff members enjoyed access that they normally would not enjoy. IT managers made presentations to the executive; some briefed the minister; some became known on a first-name basis by Administrator Garvey at the FAA; and some had large staff and budgets. In almost all cases these staff had never enjoyed that degree of exposure to the executive level. In short, most agreed that Y2K was good for their careers. Some also enjoyed more discretion than the templates would suggest. While most auditors acted as an independent check on government departments and agencies and in so doing—to a degree—undermined their authority, some government departments, particularly the BLS, did push back when auditors made claims about some systems with which the agency did not agree.

Finally, even though the United Kingdom's Science and Technology Committee report acted as an effective summary of industry's position and in fact Government adopted many of its recommendations, the House of Commons seems to have had little impact on developing the 'rules' for Y2K. The Science and Technology report was never mentioned by any interview subject, nor did the respondents mention any members of the opposition or indeed any members of parliament that were not in government. The NAO was actively involved but seemed to cooperate more with the Cabinet Office than the House of Commons. The Congress and the GAO, on the other hand, were frequently referred to by interview subjects in the United States. The FAA, in particular, felt considerable pressure from the GAO and in fact eventually resorted to hiring a member of the GAO's staff in order to help meet GAO standards and to improve relations with the GAO (INT 29).

### Sectors within the IT industry[14]

In this book I have referred to the IT industry as convenient shorthand for the sector but in fact the IT industry is extremely diverse. The industry includes providers of software, hardware, IT maintenance and consulting services, among others. The Information Technology Association of America (ITAA), for instance, has over 11,000 direct and affiliate members in the United States alone.[15] Moreover, the IT industry is not necessarily separate and distinct from non-IT sectors. As noted in this research, many organizations depended considerably on IT but many of these organizations have internal IT staff. In addition, several organizations contract not with one but sometimes several IT companies for different services. The diversity, complexity and volume make it difficult to refer to a monolithic 'IT Industry'. With respect to Y2K, some were winners and some were losers; some were neither and some were both.

In their book on management consultants, Micklethwait and Woolridge draw two conclusions: one is that the industry is potentially lucrative and two, the field of management theory can be a 'mishmash... where the books of tenured professors rub shoulders with those of out-and-out charlatans' (1996, 366). Certainly in the mid-1990s consulting was lucrative, and expanding. Gartner, one of the leading Y2K consultancies, observed: 'The management consultancy business generated $11billion worth of fees in 1994; it is on course to bring in $21billion in 1999... more than half of today's leading consulting firms did not exist five yeas ago' (cited in Micklethwait and Woolridge, 1996, 3).

The Y2K story entered popular culture *via* management gurus such as Peter de Jager, Ed Yardeni and Ed Yourdon, as well as via IT research companies such as Gartner Group, Capgemini, Data Quest, International Data Corporation (IDC), Giga Information, Jupiter and Forrester Research. De Jager was the first with his article in *Computerworld* in 1993 but Gartner, followed by Capgemini, were among the most influential. Gartner had 9100 different clients, drawn from the public and private sector, both within the United States and overseas (Feder, 1999a). Gartner sold Y2K-related consultancy to its clients for '4 to 5 figure fees'. Approximately 15 per cent of its clients were IT companies but the rest were largely government departments and various Fortune 500 companies.

Gartner as well as other IT consulting companies had considerable impact in articulating the rules of engagement with Y2K. Bruce Hall from Gartner testified to Congress: 'we must accept that risk exists in *any* technology that was ever programmed by a human, examine such technology for possible failures, and form remediation strategies' (Hall, 1997; original emphasis). Gartner increased its Y2K research capacity after it conducted its initial research and estimated that Y2K was potentially a $300–600 billion problem in the United States alone, based on very crude calculations. Quotations, such as the one noted above, helped to make Gartner (as well as other IT research consultancies) popular sources for the press. As noted in the previous chapter, 28 per cent of articles in *The NY Times* and *The Wall Street Journal* and 24 per cent of articles in *The FT* and *The Times* (of London) cited a source from the IT industry. Many were from the consulting end of the sector. They actively characterized the bug, its consequences and the necessary strategies for risk mitigation.

Gartner did not generate as much revenue as one might expect for the leading Y2K consultancy. Indeed, it could be seen as a *lost leader* to secure more clients. By 1999 Y2K services generated a mere $8 million for the company, which represented about 1 per cent of revenue. Similarly, its eight Y2K researchers represented about 1 per cent of its overall staff and research activities (Feder, 1999a). Capgemini, which has a client group similar to that of Gartner, offered similar Y2K services and earned 6–7 per cent of its revenue from these services at the time.

Partly as a result of the characterization of Y2K by Gartner, Capgemini and others, several sectors experienced increased demand for their services. In fact between 1996 and 1998 numerous Y2K consulting services emerged to seize the opportunities that Y2K offered. Without legal regulation and with such considerable uncertainty and an apparent shortage of IT labour, some outfits sensed financial opportunity. Some government organizations resorted to these specialists, particularly in 1998 when anxiety was high (NAO, 1999b, 35). The arguments associated with dependence, interdependence and embedded systems, coupled with the notion that no one could predict where the failure would emerge from, allowed IT consultants to argue that every system had to be checked, otherwize, the organization could never be sure. (This message was echoed by the Cabinet Office and OMB reporting requirements.) In one example of promotional material, a company that specialized in Y2K services cited evidence of large failure rates in tests it carried out on over 4000 computers. The material fails to indicate the model and year of purchase. Yet it concludes that all systems must be tested.[16] They frequently specialized in Y2K audits. Two such companies, Impact and PA Consulting, for instance, each audited five different sectors for the UK/NIF.

The other beneficiaries of Y2K were large IT service providers. Large IT service providers in the UK government include, for instance, EDS, ICL and Siemens.[17] These organizations were not necessarily at the forefront of Y2K. In fact sometimes Y2K was brought to their attention by their clients. These organizations were often on long-term contracts with organizations, they knew the organizations' systems very well, and rarely had to compete for Y2K contracts. Indeed, most organizations chose to use existing IT resources for Y2K rather than bring in new people (Taylor, 1999). Y2K tasks were simply added onto their workload and they were paid a supplement. In government, the duration of the outsourced contracts reinforced the influence of these suppliers. In the early phases of contracting out in the United Kingdom, for instance, the average life of a contract had been five years; but latterly, IT suppliers often secured 7, 10 and even 15-year contracts (Dunleavy *et al.*, 2001). Many of these companies did well generally during the booming IT years of the mid- to late 1990s. Y2K work may have been lucrative but it was only one of a number of initiatives that were occurring concurrently (INT 30; 34). That noted, these organizations were still influential in Y2K operations; they frequently devised the methodology that their clients adopted (INT 65; INT 66).

Hardware providers also benefited to a degree. Large hardware providers include IBM,[18] Hewlett Packard, Sun Microsystems, Dell and Unysis.

Many large organizations replace their hardware at regular intervals. The anxiety over Y2K resulted in short-term increased demand. It also allowed these organizations to sell new products and upgrades.

But were the rules devised strictly for the *gain* of the IT industry? Despite an overwhelmingly popular consensus that the IT industry benefited from Y2K,[19] in some cases the joy was short-lived and in fact posed a threat to their longer-term corporate strategies. Before I outline some of the problems with which the IT industry had to deal I will reinforce the point that the IT industry generally was booming. In an interview with a journalist in Silicon Valley, he noted that in the run up to year 2000, Y2K was a relatively small story compared to the other issues—such as the expansion of the dot-coms and e-commerce. Y2K was seen as a 'tidying up issue' that was not all that interesting (INT 31). Indeed, while the IT sector was lucrative at the time, Y2K-related work was only a small portion of most IT budgets. (Table 6.8 below.) While one might dismiss such a claim based on the volume of coverage in the newspapers cited in the last chapter, note that the coverage spans the entire US and UK economies as well as includes numerous international stories. That does not mean it was viewed with the same energy in Silicon Valley, arguably the heart of the IT industry.

If an organization wanted to profit specifically from Y2K it had to start early—by 1996. By 1998 the work and share value of these organizations peaked because most of the work had been done or at least had been contracted for. If these companies had no other specialization to offer other than Y2K-related services, then they were in trouble by mid-1998 (Auerbach, 1999; INT 52). Many of these organizations failed even to survive (INT 52). Table 6.9 lists some of the changes that occurred to some higher profile consulting organizations that had specialized in Y2K services.

Even the large IT service providers had Y2K-related anxieties of their own. Some bigger suppliers who had ongoing relationships with their clients argued that they were between a rock and a hard place. On the

*Table 6.8* Total IT and Y2K expenditures in the United States, 1996–2001

| US Economy | Expenditure ($ M) |
|---|---|
| Total IT | 4,690,769.2 |
| Year 2000 | 121,960.0 |
| Share % | 2.6 |

*Source*: Internal Data Corporation.

one hand, they did not want to be seen as 'gouging' the clients. On the other hand, as the custodians of the systems, the IT service providers felt they would be blamed if their clients' systems failed on 1 January 2000. Relatedly, if significant systems did fail, their clients might go out of business. Any of these scenarios—gouging, blaming or bankruptcy—could result in lost business post 1 January 2000 for the IT service providers. Hence, inventory, fix, test and audit represented a convenient confluence of interests between the IT service providers and clients. Ultimately, most IT service providers argue they offered the service the clients wanted.

The larger IT organizations also had their own in-house Y2K operation to worry about. These organizations are largely dependent on IT in their own right and some were concerned that their own systems would fail. Interviews with Y2K coordinators for internal operations for such companies revealed that they had similar problems as government—variety; complexity; poor documentation; organizational inertia; institutional conflict (INT 33; 46). Cutting-edge IT firms are no more interested than any one else in going through a backlog of old systems—those they use as well as those they sell—and checking for an obscure (and rather dull) bug. Indeed, some organizations had to go back to systems designed ten years earlier to investigate Y2K vulnerabilities. Moreover, their pay structure often was slow to reward it. As one interview subject noted, they (that is, divisional heads) are

*Table 6.9* Changes to some higher profile consulting organizations that had specialized in Y2K

| Organization | Y2K-related Revenue during Peak of Y2K Operations (c. 1998) | Y2K-related revenue in 1999* | Change in share value between 1998 and 1999 |
|---|---|---|---|
| Cambridge Technology Partners | Not available | Not available | **(77%) |
| Computer Horizons Corporation | 33% | 10% | ***(75%) |
| Gartner | Not available | $8 M | **(55%) |
| IMRglobal | 50% | 30% | ***(60%) |
| Keane Inc | 37% | 20% | ***(55%) |
| Peoplesoft | Not available | Not available | **(69%) |

*Forecasts printed in the *Wall Street Journal* in 11 March 1999 (Auerbach, 1999) and 17 September 1999 (McGough, 1999).
**Share value cited in 17 September WSJ (Mc Gough, 1999).
***Change in share value between April 1998 and March 1999.

paid to get new products to market, not fix old systems (INT 33). Indeed, until Y2K was considered lucrative—which it was *not* before 1997—then many in the industry were not interested. Moreover, amalgamations that had occurred in the industry throughout the 1990s created barriers between new divisions that slowed such company-wide initiatives (for example, developing a standard definition of compliance and meeting deadlines). Their problems and solutions were similar to those of government; in fact, they often sold to government the methodology they had developed for themselves.

Y2K also had the effect of perverting purchasing patterns. Organizations accelerated their purchases to avoid problems on 1 January 2000, but this change did not necessarily increase demand but merely shifted it. As a result of a purchasing freeze at the end of 1999 to avoid the introduction of new bugs after Y2K-related remediation had occurred, hardware demand plummeted (Caffrey, 1999). Xerox Corporation, NCR Corporation, Unisys Corporation and Lexmark International Group all attributed lost earnings in the fourth quarter 1999 to a slow-down in IT purchases caused by Y2K-related purchasing freezes (Bulkeley and Hamilton, 1999). IBM took the biggest hit. IBM stock valuation dropped by $29 billion in one day amid speculation about drop in demand in hardware sales (Bulkeley and Hamilton, 1999). No doubt some used Y2K as an excuse for poor income performance, and not all hardware providers did this badly. Sun Microsystems, for instance, had a strong Internet-related clientele, and its revenue grew by 25 per cent in the last quarter of 1999 (Bulkeley and Hamilton, 1999). More importantly, however, the drop in demand marked the first inkling of a malaise that would set-in on the IT industry. Indeed, after all of the downsizing, upgrading and rationalizing of IT occurred by early 2001 (projects often spurred on by Y2K programmes), the industry plummeted. All interview subjects from the IT industry agreed that Y2K was partly responsible for the collapse in the IT market in 2001, which the industry suffered from for years.

It has also been noted that Y2K was the point at which 'off-shoring' IT work occurred. With a shortage of IT labour many organizations, particularly in the United States, contracted with off-shore outfits, particularly in India. These off-shore companies used the opportunity to impress large American clients and subsequently negotiated new, non-Y2K-related IT contracts for themselves post 1 January 2000 (Merchant, 2000). This shift may have increased overall economic productivity in the United States but it may also have hurt some domestic IT firms due to lost contracts.[20]

Finally, Y2K—because of the eventual lack of drama—risked *and risks* damaging some organizations' and people's reputations. In fact, taking Y2K too seriously—by the end of 1999—became a source of material for stand-up comics (Feder, 1999c). As a consequence, many distanced themselves from their pre-1 January 2000 Y2K claims shortly after the New Year. Most of the high profile consultants that commented regularly on Y2K—in the popular press, trade publications as well as the speaking circuit, for instance—removed shortly after January 1, 2000, most if not all references to Y2K from their websites. The UN's Y2K Coordinator (a former Director of IT at OMB) described Y2K as the best experience of his life. It included preparing and overseeing a response to potential problems in the global infrastructure. But he says now he cannot even put the job on his CV. By 2 January 2000, he stated, 'Y2K was bullshit' (INT 56). IT consultancies also protect their Y2K legacies. Typically the archived reports are not made public. After an interview, one IT consultancy sent me briefing material it prepared that tended to downplay the problem (though clearly this material was in the minority of its publications at the time).

In sum, no one sector dominated Y2K. Power was diffuse and in some cases temporary. That noted, some trends emerged. A growing IT industry in the United States, for instance, identified and largely characterized Y2K and related risk mitigation strategies. Within the infrastructure as a whole, however, we see a vast array of trade-offs and game-playing between government, regulators, trade associations and industry, which largely favoured the dominant players at the sectoral level but also trusted that these players would deliver a stable, continuous service in their respective sectors when the time came.

## To what extent can pluralism and corporatism explain the governments' respective management of Y2K?

Much of the US and UK governments' responses to Y2K can be viewed through conventional pluralist and corporatist approaches. In the United States in the early stages there were no existing institutional arrangements to integrate or prioritize interests on this subject. Responses came via a large number of trade organizations testifying on behalf of massive memberships before different congressional committees. Even when the executive did intervene in the private sector's approach to Y2K, the US government was much more tepid than the UK government, despite both countries' sectors being (more or less) equally interdependent in their own right. John Koskinen noted that he deliberately created as few new institutional arrangements as possible when he arrived at the President's Council. He used the existing partnerships for information, including trade and sector level associations, which did not disrupt existing dynamics within each sector. The US/WG reports covered a large number of organizations but it did not dictate the terms of any sector's compliance. Indeed, the US government reports, while inconclusive, seem more accurate. By virtue of the technical nature of the problem, there was always a degree of uncertainty and the US reports reflect that uncertainty more effectively. No sectors identified in the US/WG were given special treatment. So far, so pluralist.

The UK/NIF exemplifies a more cohesive approach than that achieved in the United States. The select group in the UK/NIF all worked under one strategy with a standardized approach, at least at a high level. To a certain extent, this intervention represents a degree of penetration into the private sector that the US government did not achieve nor pursue. This degree of penetration served to influence strategy but did not wholly undermine the authority of the participating organizations. These organizations volunteered to participate in the UK/NIF but largely on their own terms: there would be no company-specific information made public; they would agree amongst themselves what constituted Y2K compliance; and they would agree the audit results with the auditors before they were submitted to the regulator. In some respects the audits conveyed merely the illusion of control (Power, 2004, 10). So far, so corporatist.

Despite these overarching pluralist/corporatist tendencies, neither government fits the conventional mould perfectly. The US Government still privileged the few over the many when it established the Special Advisors Group at the White House, invited selected corporations to participate in the US/WG roundtables and even signed the Y2K Act and in so doing empowered large organizations to help to manage the smaller organizations in their own sectors. Moreover, even if the US government did not dictate the terms of Y2K compliance across the infrastructure there was no doubt considerable standardization across sectors even if it were enforced only at the sector level. Similarly, the UK Government cast its net much wider than the UK/NIF; it spent considerable effort on its SME outreach through Action 2000, Taskforce 2000 and the DTI. Similarly, circumstances dictated special access. The UK government had to ensure critical services throughout the United Kingdom, which resulted, for example, in greater time spent on small food distributors because they serviced remote locations. In short, it was a multi-pronged approach. They employed a practical response to what seemed like a virtually unmanageable situation.

The Networks approach helps to bring into focus elements that are neglected by the conventional approaches. To start, most requests that were made of government by industry reflect the uncertainty *between* sectors. This grey area became the domain of the Y2K Issue Network. Industry wanted government to help ensure stability across the infrastructure without government necessarily intervening too directly in industry operations. Resource dependency was high but cross-sectoral integration was low. It was therefore in most organizations' best interest to gain some degree of reassurance that there was going to be continuity of service across the infrastructure. This problem went beyond the power of any one organization or sector and therefore the government was in an ideal position to help to coordinate a loosely defined Y2K Network that was largely interested in maintaining stability during the change-over.

But tradeoffs were necessary for this reassurance. In short, it was not going to work unless everyone played. In both countries there was pressure to participate in sector-wide activities. The UK/NIF and US/WG existed so that sectors could demonstrate compliance to other sectors as well as to the government and to the public. The US and the UK governments may have enacted different strategies but they nevertheless had the same goal: to report on and create pressure for 'readiness' across their respective infrastructures. While the sector-level arrangements were somewhat flexible, in the main, organizations were pressured to participate and ensure they were compliant. But the relationship between sectors might be thought of as co-dependence rather than interdependence. No sector, for instance, dictated the terms by which another sector would be judged compliant. The devil was in the detail and that detail was determined at the sector level in both countries.

If the details were standardized across sectors, however, then the standardization was largely influenced by the IT sector, or what might be thought of as an IT Network. The IT Network was not necessarily bound by institutions, sectors or even national boundaries. Nor was it confined to the public sector or private sector. It was pervasive, stable and loosely integrated within organizations (for example, the IT department) and across organizations (for example, consultants, IT hardware and software suppliers). Indeed, the government was not the only one that could move between traditional sectors, such as those identified in the UK/NIF and the US/WG; so too could those in IT. While no one dominated the Y2K story, the IT Network had the most sway. Among Rhodes's features of power dependence (for example, dependence, exchange, dominant coalition and rules), those in IT became the most powerful for a short time in the early period when the problem was defined and the methods and tools were developed, which essentially set the tone for the run-up to Y2K.

Finally, both governments seem to have remarkably similar approaches within government departments and agencies. With both governments demanding so much of the organizations that were directly under their control, they satisfy Marsh and Rhodes's expectations—more control over less. This similarity in their approach (and achievement) challenges assertions that these bureaucracies differ in their cohesiveness (Page, 1992, 81, 209). Moreover, it suggests that if there is a convergence between the style of the two bureaucracies it is not that the UK model is becoming more like the US model (for example, segmented and conflictual), as some NPM advocates would suggest, but rather that under circumstances like Y2K, the US bureaucracy acts more like a traditional UK model (for example, cohesive and closed) (Simon, 1999).

In sum, in the face of uncertainty and with the limited time frame the governments hurried to the existing institutional arrangements across the infrastructure by way of coordinating a relatively quick response in a complex dynamic. It was practical to do so. The UK approach was largely corporatist; the US was largely pluralist. That noted, the dynamics surrounding Y2K were simply more complex than *one* theoretical lens— or practical approach—can accommodate. The low degree of integration coupled with the vast membership and high degree of resource dependence created a Y2K Issue Network in which organizations made concessions to public reporting and transparency in order to guarantee a degree of stability across the infrastructure. The Networks approach helps to elucidate the information-coordination role between sectors that both governments played, the *intra*-sector level pressures that occur as a result of *inter*-sector level problems and the emergence of a loosely integrated yet influential IT Network.

Despite its limitations, the concept of the Issue Network brings into focus a manner of understanding Y2K and using it to advance our understanding of risk management and critical infrastructure protection. Rhodes (1981), Marsh and Rhodes (1992), Hood *et al.* (2001) and Heclo (1978) emphasize stability. Yet by its nature the Issue Network is rather ephemeral; it emerges to deal with a specific issue and disperses or lays dormant until similar conditions re-energize the network. If one views Y2K through a networks lens one starts to see how it re-emerges in current risk management and infrastructure protection practices, a point explored in the concluding chapter.

# 7
# Conclusion

> We find no villains in the federal government's officials and
> advisers then and think that anyone (ourselves included) might
> have done as they did—but we hope not twice
> —R. Neustadt and H. Fineberg (1983) *The Epidemic that
> Never Was: Policy-Making and the Swine Flu Affair*[1]

This book sought to accomplish two tasks. First, the book applied the
Hood *et al.* (2001) 'Risk Regulation Regime' framework to examine
comparatively the US and the UK governments' size, structure and
style of management of Y2K according to a cybernetic view of control,
drawing on specific examples from four government agencies, two
from the United States and two from the United Kingdom. Second,
the book used the same framework to test the extent to which such
management can be understood as responses to one (or a combination)
of three pressures: *viz.* those arising from the market, the public and
organized interests. The data concerning Y2K come mostly from official
UK/US government sources, in-depth, semi-structured interviews and
newspaper articles.

## Did context shape the US and the UK governments' management of Y2K?

The Market Failure Hypothesis (MFH) explains little of the executives'
interventions into the work of government departments and agencies.
The 'nuclear option' that the governments adopted operationalized a
step-by-step process that tried to identify, segment and eliminate risk.
There was almost no effort to manage or prioritize the work according
to the degree of risk exposure. One would have to accept a view of

systems complexity akin to that of Luhmann or Perrow[2] to adopt such a precautionary approach, which virtually treated every system as the same and necessarily absorbed considerable resources.

The MFH explains to a degree both governments' approaches to the infrastructure. Both governments helped to reduce information gaps by creating inter-sectoral fora that enabled information-sharing between numerous critical yet poorly integrated sectors. I note a few caveats, however. First, the United Kingdom's interventionist approach was likely unnecessarily ambitious as the US government's 'lighter touch' approach demonstrates. Second, in some respects, while both governments intervened by raising awareness, by not formally setting national standards they allowed market pressures to encourage organizations towards full compliance. This dynamic created market failures; it favoured larger organizations that pressured SMEs in the supply chain to carry out sometimes unnecessary Y2K audits.

With respect to the Opinion-Responsive Hypothesis (ORH) both governments were greatly concerned with the public although their actions suggest they tried to shape public opinion rather than follow it. At first, their strategies suggest they tried to align public perceptions with the pre-eminent positions of Y2K experts, which advocated a robust response to the bug. Latterly, however, governments 'talked down' the risk in the hope of reducing public anxiety and the havoc that a public overreaction could potentially cause. There are two noteworthy implications. First, both governments' operations were content that their operations not be aligned with public opinion by the end of 1999, lest it lead to public overreaction. Second, it seems more the case that management shaped context, that is, government management deliberately tried to provoke a certain reaction within the public in order to meet its objectives.

With respect to the Interests Group Hypothesis (IGH), certainly some government units gamed the system to their advantage, but overwhelmingly the CO and OMB dominated the Y2K strategy within government. It was largely a top-down process. That noted, the central offices were pressured and influenced by the national auditors, IT consultants, and in the United States, Congress. It is difficult to isolate the executive from these sources when attempting to determine cause.

Both governments' arrangements for the infrastructure were deeply rooted in the existing institutional arrangements. The UK government's reaction can largely be explained by a corporatist model while the US approach is better viewed through a pluralist lens. That noted, these conventional lenses fail to elucidate some important exceptions, in particular the inter-sectoral role the governments played in managing Y2K and the role of the IT industry, which emerged as a loosely integrated network that acted within and across sectors. Its ability to fly under the radar is perhaps best captured by the fact that the UK/NIF did not even include the IT industry explicitly in its membership.

Table 7.1 below summarizes the extent to which the hypotheses help to explain the governments' reactions to Y2K. There are a number of constraints that complicate any bottom-line scoring system, which I noted in Chapter 3. Nevertheless, in an effort to summarize, I provide an *impressionistic* scoring system to guide the reader.

In sum, among the hypotheses presented in the framework the impact of the existing institutional interests was likely the strongest influence on how the governments managed Y2K. That noted, to look at such a scale and come away with the impression that *it was the interests that did it* would be a very limiting view—both theoretically and practically—of what happened in the run-up to Y2K. The governments' responses can be viewed partly as practical given the perceived nature of the problem and the limited time available. I will move on to the limitations of the framework and then to a discussion about what this research can contribute to the theory and practice of IT risk management in government, including aspects of the case that the framework does not necessarily help to elucidate, and the CIP debate more generally.

*Table 7.1* The extent to which the three hypotheses help to explain the governments' management of Y2K

|  |  | UK | Scoring | US | Scoring |
|---|---|---|---|---|---|
| MFH | Departments | *Weak/Moderate* | 2 | *Weak/Moderate* | 2 |
|  | Infrastructure | *Moderate* | 3 | *Moderate/Strong* | 4 |
|  |  | *Total* | 5 |  | 6 |
| ORH | Departments | *Weak/Moderate* | 2 | *Moderate* | 3 |
|  | Infrastructure | *Moderate* | 3 | *Weak/Moderate* | 2 |
|  |  | *Total* | 5 |  | 5 |
| IGH | Departments* | *Strong* | 4 | *Strong* | 4 |
|  | Infrastructure | *Moderate/Strong* *(via a corporatist model)* | 3.5 | *Moderate/Strong* *(via a pluralist model)* | 3.5 |
|  |  | *Total* | 7.5 |  | 7.5 |

*Notes: Scoring System*: Weak: 1; Weak/Moderate: 2; Moderate: 3; Moderate/Strong: 4; Strong: 5
*Assumes the Executive is the dominant interest.

## The utility of the framework

References to the key elements of the framework came up frequently during interviews as well as in document analysis exercises, which led me to selecting the framework.

It was largely adaptable and a useful heuristic device, particularly given that the Y2K case has been thus far unexplored. The framework loaned itself to an inductive approach; it sign-posted several entry-points for investigation and the cybernetic concept of control and the notions of Size, Structure and Style helped to keep the concept of management sufficiently broad.

There were, however, practical problems with the application of the tools. To start, I will simply recall those limitations that I noted in Chapter 3. First, despite my summary noted above, the Size, Structure and Style categories do not lend themselves easily to aggregation and therefore a definitive bottom line comparison between countries or even organizations is likely to remain elusive. Second, the framework draws particular attention to public/private relationships, which risks marginalizing other important relationships, such as that of the IT Network, which cannot be defined strictly in public/private terms. Third, the features of the cybernetic control system overlap significantly in practice. Information-gathering can be standard setting; context is imperative to understanding how the terms are interpreted. Finally, when information-gathering and standard setting are self-reinforcing processes, it is difficult to know what information continues to fly under the radar.

Perhaps more critical, however, is the question, to what extent did these organizing concepts add to our understanding of the outcome? For instance, one might describe aspects of the style of management between the FAA and the BLS as very different. In the former case the staff was championed through the process; they were invigorated by the Executive[3]. In the latter case the staff was undermined and demoralized. Ultimately, however, the agencies approached the problem in the same way and achieved the same result.

It might be observed that the Size, Structure, Style and the cybernetic spectrum of control are merely a framework that describes; they are not meant to explain how the regimes react. The explanatory power comes from the sources of pressure from which the three hypotheses are derived. On this front, the framework was not wholly satisfying. While all three hypotheses helped to elucidate a far-reaching discussion of Y2K and provided useful insights into government reactions, none of the three hypotheses gives a fully satisfactory answer to the question, 'What shaped the governments' responses to Y2K?' In one sense, this is because all three pressures seemed to influence government reactions. But the problems in applying the framework go beyond this limitation. Allison and Zelikow (1999) note in their three-pronged approach[4] to the Cuban Missile Crisis, for instance, different approaches validate different data sources, interpret the same data differently and provide different causal explanations. This in and of itself diminishes the likelihood of identifying a single, causal relationship. But in the case of the Hood *et al.* framework, the three hypotheses cannot be viewed in isolation of each other. Markets, public opinion and interests are not simply overlapping; they are interactive. One might say, for instance, that Y2K had an impact on interests, which in turn had an impact on the public, which in turn had an impact on interests. A few iterations of such a dynamic and suddenly it becomes extremely difficult to pinpoint a causal hypothesis.

One must also consider the prospect that the framework has not identified the appropriate contextual elements. I offer two from this research, both drawn from technology. The first is meant simply as an elaboration of the technical nature of the risk, as per the MFH, and concerns the alignment between technologies and reporting methods and processes. Some technologies were spread across programme areas; often several technologies existed within one programme area. Some technologies were new; some were old. There were numerous technological interfaces. This setting required more players, more specialists, more coordination and more testing. This degree of complexity made it difficult to get a reliable snapshot of Y2K compliance; it also slowed the verification process considerably. More importantly it frustrated bureaucratic reporting mechanisms that were aligned with conventional programme areas, not technology spans.[5] This disalignment is one reason why, despite the seismic effort, Y2K compliance was still an uncertainty at the central agencies until the fourth quarter of 1999.

Second, the technical nature of the risk was critical to understanding Y2K but in fact there was little consensus over what the technical nature of the risk was. One hundred people in a room could have 100 different opinions. Some saw it as negligible; others saw it as having the potential to bring down the national infrastructure. Given the absence of concrete information, *interpreting complexity and/or uncertainty* was a critical feature in the context in which Y2K occurred. While some proffered high-level numerical estimates, most agreed that the risk was not and could not really be known until 1 January 2000. Because there could *not* be a practice-run before hand, policy-makers decided actions had to be taken about how to proceed.[6] These decisions were not made solely on the basis of the technical nature of the risk but rather on the basis of other criteria, such as political calculations and negotiations, as captured in the Hood *et al.* framework, and also in people's views of risk and technology (to be explored below).

### Implications for theory: when management shapes context, or the tail that wagged the dog

This research suggests two noteworthy outcomes for the theory of comparative public administration and IT risk management. First, contrary to Hood *et al.'s* findings, the trend was 'convergence' in government approaches to the problem rather than 'variety'. As Table 7.1 above suggests, while there were slight differences in the manner in which the two governments approached the problem within departments and agencies they were in fact remarkably similar.

What is perhaps ironic is the shape that the convergence seems to have taken. While public management reforms in the 1990s might suggest that the UK civil service was becoming more like a US model, in which initiatives such as Next Steps Agencies and the Citizens Charter might have created a more fragmented and confrontational civil service, it is remarkable how the overarching trend would suggest the opposite. That is, the US Civil Service under the threat of Y2K became more like what one would think of as a *traditional* UK Civil Service[7]—more cohesive and orderly; top-down and disciplined. Certainly there was still fragmentation due to out-sourcing; and trends such as report cards and traffic light reporting systems all had some impact. But the overall trend was one of closing ranks and getting the job done. The next point puts forward a proposal that can help shed some light on this convergence.

Second and relatedly, just as the three hypotheses are difficult to distinguish, neither can one separate context from management so readily. As noted in the ORH chapter, government interventions helped to shape public reaction, which in turn shaped government interventions. The technical nature of the risk was also subject to this dynamic. Inventory, fix and test echo standard IT risk management practices. In fact such an approach is a logical extension of PRINCE2. It embeds an objective view of risk in which the environment is assumed to be controllable, risk is understood to be negative and risk elimination is included in a list of possible outcomes. The strategies so often deployed by departments implied such a view of risk.

But these strategies expose a contradiction in government between IT management and risk management. While the IT tools conveyed a systematic and orderly approach to risk elimination, the government simultaneously conveyed complex notions of risk, in which risk could not fully be understood or controlled.

The IT risk management strategies had to be interpreted within this context also. Hence, the application of IT processes and tools in government departments and agencies unleashed a form of *organized pandemonium*, which included orderly and systematic but expansive reporting requirements and exhaustive risk elimination drills, replete with multiple layers of overlap. Outside of the immediate control of government departments and agencies, where government control was understood as precarious, the government as well as numerous key suppliers across the infrastructure forged tenuous relationships, or *paranoid partnerships*, out of necessity, in which simultaneously, everyone was in it together while also being completely on their own. Figure 7.1 depicts the results of mixing the RAP-based IT risk management tools with the different risk contexts. Departments and agencies represent a 'controlled environment' and the national infrastructure and the supply chain represent a 'fluid environment'.

Y2K represents a clash, if you will, of competing rationales. Y2K coincided with a broadening view of risk, which accommodated complexity and which had started to infiltrate institutional thinking. For this view, Y2K was a 'crisis', which helped in effect to move Y2K to what Page (2001) calls 'high politics'. Such a move implicated a much broader audience of non-specialists and in so doing a multiplicity of views in a risk-averse environment. The move to high politics made 'risk elimination' an enviable strategy; but it also discounted any measured

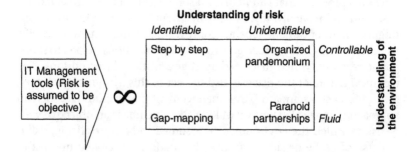

*Figure 7.1*  IT risk management tools applied to different risk contexts: when the goal is stability

approach, including for instance, cost-benefit analyses or probability risk assessments. In a sense, IT risk management tools were ill-equipped to deal with this dynamic. They were being applied in the hunt for a pervasive risk in an uncontrolled environment without the restraints of CBAs or PRAs, the devices, which would normally prevent the exercises from going overboard. Indeed, the only limitation on efforts came by way of the fixed deadline.

## Practical implications for public management

Y2K offers many lessons for government on how to manage IT and the national infrastructure more effectively. I have grouped them into three categories: IT Risk Management, IT Risk Communications, and central agency relations with departments and agencies and the infrastructure, here named the Centre and the Satellites.

### IT risk management

The threat from Y2K resulted partly from departments' mismanaging IT. The government did not know the risk to which it was exposed because in many cases, government IT units within departments and agencies did not know their systems sufficiently well. This poor understanding of their technology can be attributed to the fast pace of administrative and technological changes in the early to mid-1990s that made controlling the IT environment more difficult.

It can also be attributed to poor documentation and sloppy programming practices. In the run up to Y2K many departments spent much of their time on inventory; that is, just trying to find out which systems the organization had and then establishing whether or not the systems had any date-functionality. In many cases, they did not. Thus, while IT units were eventually more confident that their systems were not at risk from Y2K, they first had to undertake a lengthy and costly investigation. Future risks may not offer the luxury of three- to ten-year advanced warnings.

Similarly, IT units failed to act on a risk whose coming was at the very least strongly rumoured among them in the late 1980s and commonly known in the early 1990s. As Don Cruickshank reported to the Committee of Public Accounts, some early planning and better communication between the IT sector and business areas could have enabled the government to side-step much of the problem (PAC, 9 June 1999, 10). The earlier introduction of Y2K-friendly purchasing policies and Y2K-compliance clauses in service contracts with key IT and non-IT suppliers, for example, would have helped considerably. If the government is going to continue to pursue an intensive e-programme, as the Modernisation Agenda in the United Kingdom or the E-Government Strategy in the United States suggest, then IT must be managed as an important resource. Those responsible for IT must know which systems are critical to service delivery and how those systems work. They must also become more adept at scanning the environment for possible risks and acting in good time on those risks. Of course, there will always be a debate concerning what constitutes a risk and what constitutes a critical system; however, having ongoing monitoring would at least offer a sense of the operations environment. It should be noted that the US Government's *Federal Information Security Management Act (2002)*, which requires government departments to update Congress regularly on critical government systems and their security, is a step in this direction. However, the increasing pressure to integrate systems will continue to make systems more difficult to understand and control. Therefore, through the Act the government must attempt to learn and gather intelligence rather than simply adopt a template-obsessed process.

Moreover, the governments' planned move towards integration of technologies leaves itself open to the reincarnation of Y2K-type problems, in which complexity and interdependency arguments can result in massive risk management responses to small problems. Their approaches must not only provide the ability to isolate or segment errant code; they must also accommodate a complex view of the system as a whole. This, for instance, could lead to better contingency planning, including what to do in the event of a systems failure.

### IT risk communications

Second, there was a lack of informed debate. Y2K loaned itself to hysterical media headlines—health-care crises, aviation disasters and nuclear explosions. Many communications units found themselves on the defensive during 1997 and 1998, trying to calm the hysteria. The legislators, their committees, their auditors and the media themselves all seemed to be swept up with the alarmist headlines. Virtually none of the groups questioned more than superficially the assumptions upon which Y2K was predicated. Ultimately, sensational headlines, coupled with 'better to be safe than sorry' thinking, led to a call for a massive and centralized response in 1997 and 1998—and, as mentioned, the academic community was silent throughout the process.

A more proactive Y2K communications strategy might have at least challenged the early pessimism and thereby reduced the governments' overreactions in this period. IT can provoke considerable anxiety because, in psychometric terms, it is critical to the delivery of services that people 'dread' (for example, nuclear power). The more government depends on IT to deliver its service, the more vulnerable it is to a public that veers from one hyper-reaction to another. Hence, governments must be sensitive to this potential and must adopt effective communications with key partners, be they suppliers, private industry, the media or civil society at large, to reassure a potentially volatile public that they are managing IT effectively. The risk literature tells us that they can never win this battle outright but they can at least do a better job of recognizing technology as a potential source of anxiety.

### The centre and the satellites

Third, Y2K demonstrates the risks elected government assumes when it intercedes directly in the detail of how the public administration manages IT and its related risks. This policy/administration 'grey area' in which IT seems to reside opened up a convenient place for blame-shifting between the civil servants and the politicians (Hood *et al.*, 2001, 177). Despite departments and agencies having formal responsibility for Y2K compliance—indeed most departments and agencies claimed to 'own' their Y2K plans—cost-overruns, delays and the minutiae of the Y2K plans were frequently attributed to the pressure from the CO and OMB and the high level of political concern. Had there been a significant Y2K-related operational failure, with the various fingerprints on the Y2K plans from the CO and OMB down, it raises the question: 'Who would have accepted responsibility?'

Similarly, the 'inventory, fix, test, audit' everything approach—resulting from the governments' enthusiasm for Y2K compliance and desire to guarantee no disruption—not only resulted in a lengthy, expensive and perhaps excessive operation, it also pressured the civil service to accept the terms and conditions of their external IT providers and thereby drove up the cost of compliance. As the UK government continues to set deadlines centrally, such as getting services online by 2005, the government risks a resurgence of this problem.

Similarly, as a valuable resource, IT can represent a power struggle between central IT units and operational front lines, particularly within departments/agencies where a degree of IT expertise exists throughout the organization. Operational units downloaded software and purchased their own systems without IT units' permission or advice, simply because the opportunity was there and they wanted more autonomy. That is not to say that the proper management of IT requires greater centralization and stricter limits on users, as the post-September 11 environment seems to be suggesting. A more sophisticated understanding of IT management recognizes that this centralization/decentralization struggle can exist[8] and that a more supportive dynamic between IT units and operational front lines should be nurtured. Interestingly, neither the CO nor the OMB created feedback loops to verify the extent to which departments nor agencies were actually *finding* Y2K-related problems, for instance. This reinforced the dominant view that the OMB and the CO were not interested in how the IT professionals at the departmental level assessed the degree of exposure to risk.

On the other hand, as the governments' views of risk grow to accommodate multiple perspectives they must make an effort to deal with the tensions that are emerging at the interface of these alternative views. Clearly, the Cabinet Office, for instance, has divergent views of risk management—from the one that is advocated by the Strategy Unit that accommodates a constructed and/or structural view of risk to the one that is advocated by OGC that is much more closely aligned with the rational actor paradigm. The CO and other central agencies should coordinate better within their own office to ensure these divergent views are being addressed. This could include creating fora within the CO to deal with cross-cutting issues before directives are sent to agencies.

Furthermore, as the CO continues to increase its institutional 'risk management' in the form of the Strategy Unit and Civil Contingencies Unit, the CO risks being more intrusive in the departments' workload, which in turn risks Y2K-like excessive information-gathering (and the related behaviour modification). One of the CO aims in this area, for instance, is to 'reduce surprises'. With the multiple layers of staff between the operational front line and central agencies, one can envision operational risks being identified and squeezed out at every step; it risks becoming an exercise in risk elimination rather than risk management (a difference few interview subjects could articulate when asked). In short, such centralized risk management risks stifling the risk-taking culture that Number 10 and the CO claim they want to help generate (Cabinet Office, 2002). Centralized risk management must be sensitive to this dynamic. This implies not only that the CO committees strike a fine balance in their membership between departmental staff and the CO staff; it also implies departmental operational front line staff must also be involved in risk management decisions.

That noted, while both governments' broadening their views of risk to try to include competing rationales seems a noble cause, Y2K is a sobering reminder that such efforts can be fraught with pitfalls: it can slow down progress, drive up cost and create unnecessary anxieties and conflict. As noted in the literature review, these more complex and contested views of risk is one in which *politics* determines solutions. Regardless of whether this is 'high politics' or 'everyday politics', it is a dynamic in which power, persuasion and influence are key, and where there will always be some measure of discontent.

With respect to the infrastructure, the governments might seek to normalize critical relationships that had hitherto been ignored. Clearly there was an information-gulf between sectors. Opportunists selling over-the-top business continuity plans that resulted in unnecessary duplication sometimes filled this gap. Once the governments started creating fora for organizations to share information they quickly learned that things were not as bad as they had thought. Many organizations were taking steps to ensure compliance; also best practices started to be shared among participants. That noted, if these fora intend to set standards and share meaningful information rather than simply engage in public relations exercises to reassure people, then they must guard against the fora becoming an act of style over substance. Certainly the relationships will be tenuous and therefore they must be flexible in their approach to dealing with participating members. They must also be accompanied by reasonable standards and transparency otherwise they risk undermining the credibility of the entire process. At best, this can make such a process a paper tiger; at worst it can fail to achieve the stability it seeks either through operational failure or through public anxiety.

Interview subjects often bemoaned the loss of such of fora post Y2K. The 2000 petrol crisis, 9/11 and the 2003 power failure in the North Eastern United States are three high profile examples that have occurred since the disbanding of the fora at which they would have been useful. It is likely that recent acts of terrorism, however, have caused the 'wheel to be reinvented', a point to be explored in the final section.

## Context and CIP

Critical Infrastructure Protection occurs in a particular context. In many Western countries governments are taking steps to ensure the country's critical infrastructure is managed more effectively. There have been formal institutional changes, such as the creation and/or expansion of the roles and responsibilities of government departments, such as the Department of Homeland Security. There has also been a strengthening of political leadership and responsibility, as articulated in several legislative initiatives. Governments are trying to collaborate more. Formal agreements between jurisdictions and joint emergency planning exercises are more common. Governments are also looking to work more closely with the private sector, which owns the vast majority of the critical infrastructure, with an eye to managing vulnerabilities proactively.

A competitive market context does not always lend itself readily to proactive CIP. As noted in the introduction, corporate executives and their shareholders—sensitive to market pressures—are sometimes reluctant to spend on CIP because its benefits are often indeterminate. (How much money is enough/too much for risk management? One never knows.) They are also reluctant to disclose the vulnerabilities of their assets because of the risk to their organization's security, liability, share value and public image. Moreover, there is a problem with trust. Industry executives worry that sensitive information shared with government may be used (surreptitiously) for reasons other than CIP. Also, insurance coverage in this area can be expensive, and sometimes, as we have seen in this study, unreliable.

Traditionally, industries could try their luck; if they chose to take risks and failed then the market could punish them accordingly. Because organizations that manage the critical infrastructure are increasingly interdependent, however, individual decisions to under-spend on CIP and/or not disclose CIP-related information is now a risk for the entire critical infrastructure and all those who depend on it. This is indeed a market failure to which government will have to continue to respond; yet its capacity to pry potentially market-sensitive information from firms in competitive markets will always be constrained.

The market context is not the only relevant context in the discussion about CIP, however. As noted, the opinions of the media and civil society are also important. CIP media coverage often goes in fits and starts: it peaks in a crisis and then falls away. The infrastructure failures we learn about most often are spectacular failures. The insatiable appetite of 24/7 media coverage influences not only the way civil societies understand the problem but also who, if anyone, is to blame. This pressure generates considerable incentive for short-sighted reactions and blame-avoidance strategies among key stakeholders.

The role of organized interests is particularly relevant in recent CIP policy initiatives. Each country has its way of interacting with industry. UK Government interaction with industry is an elite endeavour: it is closed and selective. Exchanges between the two are underpinned by relatively high trust, cooperation and informal standards. The US government, in contrast, communicates with industry at a distance, largely through industry associations. Standards are enforced by laws or by markets. Trust is low among government, industry and citizens.[9]

Each approach will have its challenges. The closed approach to government/industry regulation in the United Kingdom, for instance, is likely to be slow to accommodate the growing importance of international supply chains, small- and medium-sized enterprises and the increasing role of pan-European institutions. This inertia will not only challenge the suitability of the UK government's arrangements it will bring into question the accuracy of government reports about the critical infrastructure as a whole. In the United States, on the other hand, one of the most powerful mechanisms the government is likely to have is not a closer relationship with industry but rather a capacity to help create *demand* for security through public relations exercises such as congressional hearings, press conferences and traffic light mechanisms (for example, red, orange, yellow, blue and green lights of Homeland Security's national threat advisory). In short, whereas the United Kingdom may have significant deficiencies in its consultation and disclosure methods, the United States will be constrained by a prescriptive legal context and a shortage of trust among key stakeholders.

No one hypothesis holds the answer to why governments respond the way they do. But by testing each hypothesis we come closer to understanding the issue in the round. We gain useful insights to the multiple pressures that potentially influence policy decisions about risk, each with its own merits and potential drawbacks.

Governments are not merely subject to these contextual pressures, however. They can apply pressure themselves. Governments not only respond to the law, through legislative bodies they make the law. Most Western governments have sought and received the backing of their national legislatures as they have expanded their intentions for national CIP strategies. Nor do they simply react to a 24/7 media. As we have seen here, they engage civil society through the media. Indeed, their capacity to strike an appropriate balance between transparency and discretion will be important in earning popular approval in this area.

Finally, with respect to interest group interaction, the government has the capacity to add value by facilitating the exchange of information about vulnerabilities and best practices across policy areas. In most Westminster countries at least, government is the one constant in all sector level fora. Certainly some sectors will be easier to work with than others. Natural monopolies (for example, bridges or water supply), for instance, are potentially in a better position to share sensitive information across organizations than organizations in competitive, multi-organizational, multi-sectoral settings. These monopolies might be encouraged to share lessons learned with others.

Government's success will depend partly on its capacity to share or facilitate the sharing of meaningful information efficiently. This means overcoming legal constraints, yes, but also turf wars within governments and trust problems with industry. It also means ensuring that the data exchanged are compatible. This is easier said than done. Different sectors and jurisdictions have different ways of gathering information.

Voluntary fora—such as the ones most governments propose for CIP initiatives—derive their influence through persuasion, trust or membership self-interest. They are often tenuous arrangements. Information is filtered through biased industry associations. When things go wrong participants drop out. They threaten to sue if their security lapses will be disclosed. The governments risk losing their capacity to act as arbiters for the sector, knowing that by actively participating in these fora their authority diminishes as they become merely interested participants at a round-table.

Governments must be sensitive to all contextual pressures but not captured by any one in particular. Striking the balance between these inherent tensions will not be easy. By the same token the problems are not likely to go away. The Department of Homeland Security (DHS) is here to stay. It was borne of a particular context and to fulfil a need. It is now *part* of the context. Much like the creation of the Environmental Protection Agency (EPA) altered the debate about the environment, the creation of DHS will change the dialogue about domestic security and the context in which it occurs.

As stated at the outset of this book, Y2K may seem rather mundane in a post-9/11 environment. In fact, the Y2K case is an excellent springboard into research on infrastructure protection. As IT continues to offer the possibility of reduced cost and increased service, then governments will continue to invest more intelligence, resources and capabilities in it. But efforts at technological integration will increase systems complexity; institutional barriers will persist; so too will the competing views of risk within which the technology is interpreted.

Post-9/11 IT staff in US government departments and agencies are currently working on strategies to ensure their IT infrastructure is secure and can withstand major breakdowns in the infrastructure. Numerous interview subjects raised the exercise during our discussions. The *Federal Information Security Management Act (2002)* has formalized a process to track critical IT in government. The process is similar to that of Y2K and risks incurring the same problems: template-driven; managed through OMB; verified by external auditors; and submitted to Congress.

9/11 has also had an impact in the United Kingdom, of course. As noted, the institutionalized presence of risk management at the CO continues to expand in the form of the Strategy Unit and Civil Contingencies Secretariat. The reaction after the underground bombings in London on 7 July 2005 brings stark similarities with Y2K into focus. One report following the event on Channel Four News featured a service provided by Cable and Wireless in which it backs-up all electronic files in a secure (undisclosed) location for major organizations with a presence in London to ensure business continuity in the event of, for instance, the failure of the national grid. The programme featured infrastructure protection consultants (London First) arguing that the problem went beyond the major companies in London; SMEs, critical to the supply chain, were also vulnerable as were organizations located outside the capital. The programme features 'self-assessment' polls among industry leaders that reveal high anxiety. The ABI (The Association of British Insurers) was also featured and argued that existing insurance policies do not cover organizations for terrorism and that they would be pressuring the government for standards in this area (Wednesday, 4 August 2005).

This striking similarity is not actually surprising. The Regime concept foregrounds the perseverance of such systems. On the one hand this means that as events like 9/11 and the London bombings occur, the usual suspects—both inside and outside government—will emerge to characterize the events and present a plausible way forward. Markets and existing institutionalized interests will play a strong role in shaping the response. It also demonstrates the value in studying events such as Y2K. To some extent the Y2K Evangelists have been reborn as Infrastructure Protectionists. Their manner of interpreting events today will not only be similar to the manner in which they interpreted events in the run-up to Y2K, they will be filtered through the experiences they gained in the run-up to Y2K. In fact, viewed through this lens, Y2K does not seem so isolated after all. Indeed, such cases help us to understand the regimes—who the players are; their interests; the preferred institutional arrangements; the reactions of the executive, the legislature, the departments and agencies and the media; and how in the face of uncertainty proposals will be put forward to identify and control selected risks of an unknowable future.

# Appendix A
# Methodology

## Method in brief

This book is the result of a comparative case study that seeks to explain by inductive means the manner in which the US government and the government of the United Kingdom managed the risks associated with Y2K. I have endeavoured to develop a robust and balanced research design to enable comparison between countries and agency types while minimizing the impacts of extraneous variables. First, I selected agencies with comparable remits. Second, I tried to ensure parity in documentary and interview sources and perspectives between each of the two countries and agency types in question. Third, I have attempted to minimize the problems associated with merely juxtaposing the two countries' case studies, which risks failing to compare the countries sufficiently on any thematic grounds. By way of combating this problem, for instance, rather than describe the governments' reactions along national lines, I have grouped, compared and contrasted approaches by agency type (that is, executive; statistics agencies; aviation agencies) along Hood *et al.*'s categories (that is, Size, Structure, Style) as well as by the critical elements identified in each of the hypothesis chapters.

Despite the effort to balance the approach, there are five areas in which I did not achieve this balance. These methodological deficits can be attributed to the different institutional or cultural arrangements that exist in both countries, as well as to practical trade-offs that were made during the course of the research. First, I interviewed more people in the US IT industry than in the UK IT industry. This difference was partly determined by the opportunities that presented themselves to me and partly due to the fact that the American IT industry was ostensibly much more active in shaping the Y2K story and therefore its perspective was perhaps more desirable. Second, I interviewed more legislative staff from the United States than the United Kingdom. The Congress was much more active on the question of Y2K and therefore its perspective seemed more important. Third, I cite evidence from six US public opinion polls and only two UK opinion polls. Public opinion seemed to be more newsworthy in the United States compared with the United Kingdom and relatedly the National Science Foundation polled more frequently in the United States than MORI, for instance, did in the United Kingdom: thus there is simply more data to call upon. Fourth, I interviewed staff from lead departments as well as agencies in the United States whereas in the United Kingdom I only interviewed staff from the agencies. Relations with lead departments came up more frequently in interviews with agency staff in the United States, whereas in the United Kingdom, Cabinet Office was more frequently the point of reference by agency staff. Fifth, I conducted more interviews in the United Kingdom than in the United States when it came to agencies other than my primary agencies of interest. This difference is largely explicable because I was based in the United Kingdom and at first I was interviewing staff from a cross-range of agencies until I could determine the most appropriate ones for primary focus.

## The comparative method and country selection

Y2K represented arguably the *same* risk at the *same* time in both countries. As a discrete episode it has a beginning, a middle and an end. As such, at least some variables are held constant, which allows for a closer examination of the differences between the countries. In some respects, Y2K was largely an Anglo-phenomenon,[1] and therefore any universal declarations about Y2K should be treated with caution. The US and UK governments had robust reactions to Y2K relative to other countries. In fact, it might be suggested that two countries with *very different reactions to Y2K* might be better suited for comparison. Before I note the challenges to such a research project, however, let me note first that the assertion assumes that the US Government and the UK Government *had* similar reactions. This is not strictly so, and in fact, by comparing the United States and United Kingdom, we get to see how two countries that both orchestrated large reactions to Y2K can still exhibit differences in their approaches.

Returning to the other approach in which we would examine opposite cases, however, it is not as easy as one might think to determine which countries had a large-scale response to Y2K and which did not. Moreover, even if there are differences, it is unclear what observations one might be able to make about risk management in the light of the differences. I list here three problems with such an approach. First, there are challenges in determining just who had a relatively modest reaction to Y2K:

- many European bureaucracies had Y2K operations in place on a division-by-division basis but did not have a central Y2K office in place and therefore were unable to communicate the depth and breadth of their response in any coherent way;
- many Asian and Eastern European countries did not start early but built-up their response later, benefiting from the knowledge and experience of those who had started earlier;
- many governments might have had larger reactions to Y2K than many analysts had thought at the time but now these governments would probably be reluctant to admit that they had such a large reaction to Y2K, lest they be thought of as having been taken advantage of by the IT industry.

Second, among those countries that had modest reactions to Y2K, there were a host of reasons that might explain the reaction that did not always relate to their belief in the seriousness of the potential consequences of Y2K-related failures:

- many (most) countries are not as dependent on technology as the United States and United Kingdom are;
- some governments' were in deficit and did not have the money to put towards Y2K (not so in the United States at the time);
- many governments arrived relatively late to modern technology;
- many purchased and installed much of their IT by the mid- to late 1990s and in effect installed equipment that was already Y2K compliant.

Third, among those that did not apparently have a large reaction to Y2K, methodologically, it is very difficult to study such a 'non-reaction' for the following reasons:

- it is difficult to study something that did not happen: the counterfactual problem. There are no files; there were no meetings; no money was spent. I note this with experience because originally I considered (the few) US and UK agencies that had a relatively small reaction to Y2K. In short, it became very difficult, very quickly to study the case;
- relatedly I would be heavily dependent upon after-the-fact accounts, which, given that nothing really happened anywhere, might be an all-too-rosy and self-congratulatory account of how they did not fall for a 'scam' created by the IT industry.

Finally I will note that I believe a research design that compared, for example, the United States' and Russia's approach to Y2K could be developed and could be insightful. Nevertheless, many of the points raised above would still have to be addressed and would still involve a series of trade-offs.

## Agency selection

While both agencies are very dependent upon complex technology to perform their services, there are some important differences that distinguish them viewed through the lens of each of the three hypotheses. With respect to the Market Failure Hypothesis, the difference largely rests in the cost of operational failure. While both statistics agencies publish market sensitive reports (for example, those concerning inflation and employment reports) any reduction in the aviation sector in either the supply side (caused by operational failure in aviation) or the demand side (caused by loss of confidence by the travelling public) would have significant and immediate impacts on the domestic economies.

The Opinion-Responsive Hypothesis also reveals a marked difference between the agency types. While there might be some media coverage and public concern if the statistics agencies experienced operational failures, the reaction would be minor compared to that which would occur if there were operational failures in the aviation industry, again, due to its importance to the infrastructure, the regularity with which many people directly engage its services, the health and safety impli-cations and the anxieties people feel about flying. (See, for example, Slovic, 1992.)

With the Interests Hypothesis, again there is a marked difference. The FAA and the CAA are regulators. They make and enforce rules that affect the service and financial success of those they regulate. The aviation industry, in particular, is dominated by a relatively small group, (arguably) what Wilson (1980) would describe as 'Client Politics', in which the benefits are concentrated in a handful of industry members but where the costs are dispersed widely. This dynamic is particularly relevant in the United Kingdom where 15 companies represent 80 per cent of the industry in terms of passengers and cargo carried (Action, 2000, 1999). The research necessarily then brings into focus the relationship between regulators and regulated and the negotiations between the two in the run-up to 1 January 2000. The statistics agencies, on the other hand, pride themselves upon their independence. They have no dominant client group; they are even expected to be largely independent of their

*Table A.1*  Variation within agencies along key measures of the three hypotheses

| Hypothesis | Component of hypothesis | Measure | Aviation | | Statistics management | |
|---|---|---|---|---|---|---|
| | | | FAA (US) | CAA (UK) | BLS (US) | ONS (UK) |
| Market failure hypothesis | Technical nature of the problem | Dependence on technology | Med-High | Med-High | High | High |
| | Market sensitivity | Consequences to the economy in the event of operational failure | High | High | Med-low | Low |
| Opinion-responsive hypothesis | Public opinion | Anxiety generated by operational failure | High | High | Low | Low |
| | Media profile | Coverage in the event of operational failure | High | Med-High | Med-Low | Low |
| Interest group hypothesis | Interests | Interaction with external interests | High | High | Low | Low |

lead government departments in order to ensure their reports are and seem completely independent of government influence. Table A.1 below summarizes the variation between agencies by critical components of each of the three hypotheses.

## Interviews

The interviews covered a broad range of material; they also included two different styles of questioning techniques. Interviews started with semi-structured questions that reviewed the IT environment at the agency and how it has changed over the previous ten years. Questions on Y2K, in particular, focused on how it was managed: the institutional arrangements and practices that were adopted; the interview subject's impressions of the risk Y2K posed to operations; the role of external consultants and auditors; what interview subjects would do differently if given the chance. The interviews ended with more pointed agree/disagree questions that applied slightly increased pressure on interview subjects to express opinions on specific themes, for example, the effectiveness of risk management; value for money objectives. This different tack was adopted to offset problems encountered in early and pilot interviews in which interview subjects were not always forthcoming on slightly more controversial issues.

My interviews occurred in Belfast, London, Washington DC and San Francisco (Silicon Valley) in the period September 2002–May 2004. My initial interviews occurred in Belfast because of ease of access. It allowed me to develop my interview tools and narrow my list of target agencies. In total, I interviewed 68 people. Interview subjects included government IT programmers, project managers, auditors, the President's advisor on Y2K, Silicon Valley news correspondents and executives from the IT industry.

The interviews were (usually) taped and lasted about 90 minutes. I prepared detailed transcripts and e-mailed them to interview subjects to allow them to comment on their accuracy. I was rarely refused an interview. Indeed, most IT staff that worked on Y2K were quite eager to discuss it, though undoubtedly the interview results were compromised somewhat by the passage of time between the events themselves and the interviews for this research. Many key decisions about Y2K were made between 1997 and 1998. Figures A.1 and A.2 below summarize the roles of interview subjects.

## Document analysis

I reviewed 800 newspaper articles from six different newspapers. I also analysed primary and secondary government documents, including department and agency Y2K files, GAO/NAO publications and documents from the National Archives (US and UK), Library of Congress and UK Parliamentary Library. Table A.2 below gives an overview indication of the quality of access I experienced at each agency with respect to documentary evidence and interview subjects. As a guide for the reader I have scored access on a scale from zero (nil) to five (excellent). I have then averaged each agency score in the right hand column. Obviously this assessment is based on my own impressions.

## Media analysis

For the content analysis of the six newspapers and the selection of articles, I identified my sample by drawing on all those articles that appeared in the period 1 January 1997–31 December 2000 and that included in the

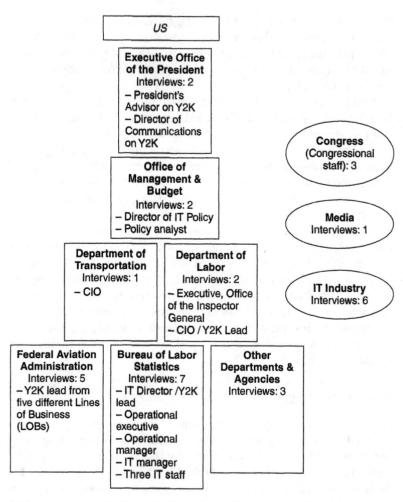

*Figure A.1*   Interviews US, sorted by institution

headline the term(s) 'Y2K', 'millennium bug', 'millennium bomb' and/or 'Year 2000 computer problem' in the six newspapers selected. The media coverage is not comprehensive. It does not include, for instance, articles that use less common terms in the title to refer to the bug. This search method limits the results particularly in the 1997 period when many terms were being used to describe Y2K. It also does not include articles that do not refer to the key search terms in the title but do refer to them in the body of the text. Searching for the reference in the body of the text provides a

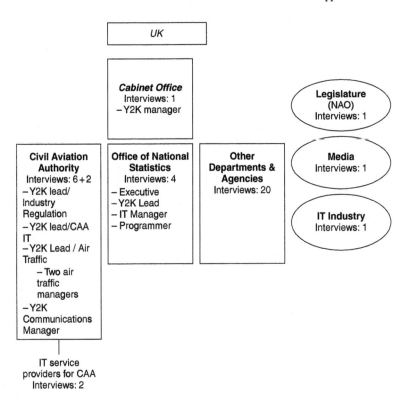

*Figure A.2*   Interviews UK, sorted by institution

considerably larger number of articles, but it includes many articles relating to the millennium that have nothing or little to do with the computer bug, or which mention the bug only in passing. (Note, 'Y2K' also referred to the New Year in a very general way). For this reason, I chose to search for articles with the key terms in the title and thereby identify articles that were primarily about the computer bug. I eliminated any articles that were clearly not about the bug. I applied the same search method to *Computerworld*. With *Government Computing*, however, I drew the articles from a sample of editions that were issued in the period 1998–2000.

For the analysis of the headlines and text, I drew on the analytical framework of Rowe, Frewer and Sjoberg (2000), which examines not only volume but also media tone and content when considering how science and technology is communicated to the public. Headlines were categorized in one of four ways: alarming; reassuring; alarming and reassuring; neither alarming nor reassuring.

Table A.2 Access to documents and staff

| Country | Agency | Access to documents | | Access to staff | | Score (out of 20) |
|---------|--------|---------------------|---------------------|-----------------|------------------|-------------------|
| | | Publicly available documents | Access to private files | Executive level | Operational staff | |
| UK | ONS | Weak (1) | Excellent (5) | Moderate (2) | Moderate (2) | 10 |
| | CAA | Moderate (2) | Strong (4) | Weak (1) | Strong (4) | 11 |
| | Cabinet Office | Strong (4) | Good (3) | Nil (0) | Good (3) | 10 |
| | NAO | Strong (4) | Good (3) | Strong (4) | Nil (0) | 11 |
| US | BLS | Good (3) | Moderate (2) | Good (3) | Excellent (5) | 13 |
| | FAA | Excellent (5) | Weak (1) | Moderate (2) | Strong (4) | 12 |
| | EOP | Strong (4) | Nil (0) | Excellent (5) | Nil (0) | 9 |
| | GAO | Excellent (5) | Nil (0) | Nil (0) | Good (3) | 8 |

*Score:* Nil (0); Weak (1); Moderate (2); Good (3); Strong (4); Excellent (5).

I reduced the impact of bias in assessments by the use of several strategies. I assessed all the articles during a short and fixed period of time (between July and September 2004) to ensure some consistency of impression. I also developed and applied a standard template to all articles. All results were stored in an Access database that I developed and maintained.

The Opinion-Responsive Hypothesis chapter includes graphs that compare the 'alarming' headlines across the papers and across time. A few methodological constraints bear noting on the content analysis of the headlines of the selected media. First, what constitutes alarming is often in the eye of the beholder. In general, I judged headlines to be alarming when they used dramatic language and implied the potential of a negative outcome. As best as possible, I tried to wear the hat of the newspapers' (relatively broad) target audience and not that of an IT specialist. While there were some difficult judgements, most headlines were categorized easily. Second, headlines are dramatic. Just because a headline is alarming it does not follow that the entire article is alarming. Third, *The Financial Times* used the term 'millennium bomb' to describe Y2K for much of its coverage in 1997 and early 1998. I chose to classify this treatment of the bug as 'alarming', and therefore whenever it appears the headline is listed as such. This decision may seem a little harsh; after the term was used for several months it no doubt lost some of its dramatic effect.

For analysis of the content of the articles, I simply counted the number of articles that referred to various key terms for this research. The key search terms were selected based on terms that related to the Hood *et al.* (2001) framework and the Rowe *et al.* (2000) framework, as well as to other conventional items that related to questions of public administration and management and were either (regularly) present in the media content or conspicuous by their absence.

Word count differences pose a challenge for reliable comparison across newspapers, and in this case, across countries. The US newspapers, on average, were much longer. Table A.3 captures the average word count per article per newspaper for the articles analysed.

This difference has a ripple effect in the analysis. Because the American articles were longer—and they tend to be longer in general, not just in the case of Y2K stories—the US papers used more sources and employed more references. The word count difference, therefore, undermines (somewhat) a direct comparison between the coverage within the respective countries, and especially when one counts and interprets the use of references, as I did.

*Table A.3* Median word count per article, sorted by newspaper and by quarter

| FT | LT | Sun | UK Average | WSJ | NYT | USAT | US Average |
|----|----|-----|-----------|-----|-----|------|-----------|
| 408 | 343 | 263 | 338 | 736 | 594 | 600 | 643 |

Note also that this analysis (like Hood *et al.'s*) did not include any TV coverage, websites or online chat lines. The reason for excluding this coverage was partly due to limited financial resources and time but also because of the ephemeral nature of these sources. Some interview subjects felt all three had considerable impact (for example, INT 49; INT 30). While TV sources were referred to the most, I will make a special note about specialized webpages, in particular. The Y2K story had several of such websites, followed eagerly by both online junkies and concerned business managers. Peter de Jager's Y2K website, for instance, received over 500,000 hits in July 1998 alone (Romano, 1998). Gary North was the author of a much used 'alternative' Y2K website. North was aY2K commentator who saw disaster and conspiracy throughout the run up to the year 2000. It seems that he had a large following and he did receive some coverage from the mainstream press. Unfortunately, the pages are no longer accessible.

# Appendix B
# List of Interviews

Legend:

| | | | |
|---|---|---|---|
| UK: | United Kingdom | CAA: | Civil Aviation Authority |
| US: | United States | NATS: | National Air Traffic Service |
| Intl: | International | FAA: | Federal Aviation Administration |
| GB: | Great Britain | DoT: | Department of Transportation |
| CS: | Civil Service | ONS: | Office of National Statistics |
| IT: | Information Technology | BLS: | Bureau of Labor Statistics |
| Ind: | Industry | DoL: | Department of Labor |
| UN: | United Nations | EDS: | Electronic Data Services |

Selected summary

| Interviews | Country | Sector | Org | Dates |
|---|---|---|---|---|
| 68 | UK: 34<br>US: 31<br>US/UK: 1<br>UK/Intl: 1<br>US/Intl: 1 | NI CS: 15<br>GB CS: 16<br>US CS: 23<br>IT Ind: 9<br>Media: 2<br>Brit/UN: 1<br>US/UN: 1<br>Health: 1 | CAA/NATS:<br>9 (includes<br>EDS staff)<br>FAA/DoT: 6<br>ONS: 4<br>BLS/DoL: 9 | All interviews<br>conducted<br>between<br>September<br>2002 and May<br>2004 |

Interviews (chronological order)

| Code | Country | Sector | Org | Job Title/<br>Descriptor | Interview Date |
|---|---|---|---|---|---|
| INT 1 | UK | NI CS | Department for<br>Social<br>Development | Deputy<br>Permanent<br>Secretary | September<br>2002 |
| INT 2 | UK | NI CS | Department<br>for Regional<br>Development | Permanent<br>Secretary | September<br>2002 |
| INT 3 | UK | NI CS | Central IT Unit | Director | September<br>2002 |
| INT 4 | UK | NI CS | Central IT Unit | Manager | 10 October<br>2002 |

(Continued)

| Code | Country | Sector | Org | Job Title/ Descriptor | Interview Date |
|---|---|---|---|---|---|
| INT 5 | UK | NI CS | NI Statistics and Research Agency | Head of Central Survey Unit | 12 February 2003 |
| INT 6 | UK | NI CS | Public Records Office NI | Information Systems Manager | 21 February 2003 |
| INT 7 | UK | NI CS | Ordnance Survey NI | Head of IT | 28 February 2003 |
| INT 8 | UK | NI CS | Department of Regional Development | Project Manager* | 5 March 2003 |
| INT 9 | UK | NI CS | Department of Regional Development | Y2K Coordinator* | 5 March 2003 |
| INT 10 | UK | NI CS | Ordnance Survey NI | Systems Manager | 6 March 2003 |
| INT 11 | UK | NI CS | Department of Finance and Personnel | Business Services Manager | 13 March 2003 |
| INT 12 | UK | NI CS | BBC NI | Manager Broadcast Services* | 20 March 2003 |
| INT 13 | UK | NI CS | BBC NI | Technology Executive and Operational Manager of Broadcast Engineering* | 20 March 2003 |
| INT 14 | UK | NI CS | BBC NI | Technology Support Executive* | 20 March 2003 |
| INT 15 | UK | British CS | ONS | Y2K Team Leader | 7 April 2003 |
| INT 16 | UK | British CS | Public Records Office | Head of IT | 8 April 2003 |
| INT 17 | UK | British CS | Radio Communications Agency | Service Management* | 9 April 2003 |

| INT 18 | UK | British CS | Radio Communications Agency | Facilities Manager* | 9 April 2003 |
|---|---|---|---|---|---|
| INT 19 | UK | British CS | Radio Communications Agency | Head of IT* | 9 April 2003 |
| INT 20 | UK | British CS | Historical Manuscripts Commission | IT Staff | 10 April 2003 |
| INT 21 | UK | British CS | ONS | Divisional Director of National Expenditure Income Division | 11 April 2003 |
| INT 22 | UK | British CS | CAA | Y2K Ind Liaison* | April 2003 |
| INT 23 | UK | British CS | CAA | Head of IT* | April 2003 |
| INT 24 | UK | British CS | CAA | Assistant Director of Communications | May 2003 |
| INT 25 | US | US CS | American Battle Monuments Commission | Director of Personnel and Administration | 15 October 2003 |
| INT 26 | US | US CS | National Archives and Records Administration | Director of IT | 16 October 2003 |
| INT 27 | US | US CS | BLS | Director of Survey Processing* | 22 October 2003 |
| INT 28 | US | US CS | BLS | Programmer/ Analyst* | 22 October 2003 |
| INT 29 | US | US CS | FAA | Y2K Lead— Research and Acquisitions | 22 October 2003 |
| INT 30 | US | IT Ind | Hardware/ Services | Y2K Lead (Executive) | 23 October 2003 |
| INT 31 | US/UK | Media | Newspaper | IT Journalist | 7 November 2003 |

(Continued)

| Code | Country | Sector | Org | Job Title/ Descriptor | Interview Date |
|---|---|---|---|---|---|
| INT 32 | US | Health Care | University of California San Francisco | Head of IT | 10 November 2003 |
| INT 33 | US | IT Ind | Hardware/ Services | Y2K Lead | 10 November 2003 |
| INT 34 | US | IT Ind | Consulting Services | Y2K Analyst | 13 November 2003 |
| INT 35 | US | US CS | BLS | Branch Chief for Current Employment Statistics Systems* | 20 November 2003 |
| INT 36 | US | US CS | BLS | Program Manager for Current Employment Statistics* | 20 November 2003 |
| INT 37 | US | US CS | BLS | Computer Special- ist—Current Employment Statistics* | 20 November 2003 |
| INT 38 | US | US CS | FAA | Y2K Lead— Public Affairs | 20 November 2003 |
| INT 39** | US | US CS | Y2K Senate Committee/ General Accounting Office | Professional Staff | 21 November 2003 |
| INT 40 | US | IT Ind | IT Service Provider | Y2K Lead for External Clients | 21 November 2003 |
| INT 41 | US | US CS | BLS | IT Manager, Intl Price Systems* | 24 November 2003 |
| INT 42 | US | US CS | BLS | Branch Chief, Intl Prices* | 24 November 2003 |
| INT 43 | US | US CS | Y2K Senate Committee/ Homeland Security | Professional Staff | 24 November 2003 |

| INT 44 | US? | US CS | FAA | Y2K Lead—Air Traffic Control | 25 November 2003 |
| INT 45 | US | US CS | FAA | Y2K Lead-Other | 26 November 2003 |
| INT 46 | US | IT Ind | IT Service Provider | Y2K Lead— Internal Operations | 26 November 2003 |
| INT 47 | US | US CS | President's Council on Y2K | Communications Director | 1 December 2003 |
| INT 48 | US | US CS | DoL | Deputy Assistant Secretary (Y2K Lead) | 3 December 2003 |
| INT 49 | US | US CS | FAA | Y2K Communications | 3 December 2003 |
| INT 50 | US | US CS | DoL | Office of the Inspector General | 4 December 2004 |
| INT 51 | US | US CS | Office of Management and Budget | Analyst | 5 December 2003 |
| INT 52 | US | Media | Newspaper | Business Journalist | 8 December 2003 |
| INT 53 | US | US CS | Subcommittee of Government Management and IT | Political Staff | 8 December 2003 |
| INT 54 | US | US CS | President's Council on Y2K | President's Advisor on Y2K | 9 December 2003 |
| INT 55 | US | US CS | DoT | CIO | 9 December 2003 |
| INT 56 | US/Intl | US CS/ UN | Office of Management and Budget/UN Y2K Cooperation Center | Director | 10 December 2003 |
| INT 57 | US | IT Ind | IT Service Provider | Y2K Lead | 11 December 2003 |

(Continued)

| Code | Country | Sector | Org | Job Title/ Descriptor | Interview Date |
|------|---------|--------|-----|------------------------|----------------|
| INT 58 | UK | IT Ind | IT Consultant | Analyst | 15 December 2003 |
| INT 59 | UK | British CS | CCTA | IT Management | 14 April 2004 |
| INT 60 | UK | British CS | National Audit Office | Director of IT | 15 April 2004 |
| INT 61 | UK | British CS | ONS | Programmer | 15 April 2004 |
| INT 62 | UK | British CS | ONS | IT Manager (Infrastructure) | 20 April 2003 |
| INT 63 | UK/Intl | British CS/UN | NATS/UN | Y2K lead at NATS/Y2K Aviation Office at UN | 21 April 2004 |
| INT 64 | UK | British CS | CAA | Manager UK Register of Civil Aircraft | 23 April 2004 |
| INT 65 | UK | IT Ind | IT Service Provider | IT Project Management* | 23 April 2004 |
| INT 66 | UK | IT Ind | IT Service Provider | IT Project Management* | 23 April 2004 |
| INT 67 | UK | British CS | National Air Traffic Control | Facilities Manager* | May 2004 |
| INT 68 | UK | British CS | National Air Traffic Control | Facilities Manager* | May 2004 |

*Notes*: *Asterix denotes that the interview was conducted with at least two interview subjects present. All other interviews were conducted one-on-one. In addition, I conducted four pilot interviews on December 2001 at the Ontario Public Service, Canada.
**GAO staff on secondment in the last 18 months to Y2K Senate Committee.

# Appendix C
# Members of US/WG on Transportation and the NIF Group in Aviation

## US/WG on Transportation

Department of Transportation (Chair)
Department of Agriculture
Department of Defense
Department of the Interior
Department of State
Department of the Treasury
Federal Trade Commission
National Aeronautics and Space Administration
US Postal Service
Air Transport Association
Aircraft Owners and Pilots Association
American Association of Airport Executives
American Association of Motor Vehicle Administrators
American Association of Port Authorities
American Association of State Highway Transportation Officials
American Petroleum Institute
American Public Transit Association
American Trucking Association
Association of American Railroads
Chamber of Shipping of America
Community Transit Association of America
Federation of International Trade Associates
Hazardous Materials Advisory Council
International Association of Chiefs of Police
International Council of Cruise Lines
ITS America
National Air Carrier Association

National Business Aviation Association
United Motorcoach Association
Transport Topics
Transportation Institute

## NIF Group in Aviation

ATC:      National Air Traffic Services
Airports: BAA Plc
          Manchester
Airlines: Air 2000
          Airtours
          KLM UK
          Britannia Airways
          British Airways
          British Midland
          Caledonian Airways
          Channel Express
          Flying Colours Airlines
          Monarch Airlines
          Virgin Atlantic Airways

# Notes

## 1 Introduction

1. See, for example, Beckett (2000b) for the UK and US Senate Special Committee on the Year 2000 Technology Problem (2000) for the United States.
2. US–Canada Power System Outage Task Force (2004), *Final Report*. Available at https://reports.energy.gov/.
3. It is a slightly modified version of the framework. The important modifications are noted primarily in this chapter; however, there are modest modifications noted throughout the book.
4. There have been some government post-mortems (see, for example, Cm 4703 (2000) and GAO (2000)) but there has been virtually no academic research on the topic. I note here my own published research that will be drawn upon throughout this book (Quigley, 2002; 2004; 2005a; 2005b; 2006; 2008).
5. (1) Attacks by dangerous dogs outside the home; (2) lung cancer caused by radon gas at home; (3) and at the workplace; (4) cancer caused by benzene from vehicle exhaust; (5) and at the workplace; (6) attacks on children by paedophiles; (7) injuries and deaths from vehicles on local roads; (8) health from pesticides in food; (9) and in water (Hood *et al.*, 37).
6. The important distinction is not in the terms but rather in the focus of investigation. I include government departments and agencies as a primary focus of investigation, whereas Hood *et al.* looked primarily to understand how the government regulated others. The term 'risk management' will be used throughout the book to describe the risk practices that were the focus of this research. Essentially, for the purpose of this book, risk regulation and risk management can be thought of as synonyms.
7. CCTA was not formally part of the Cabinet Office but worked closely on Y2K with CITU.
8. See, for example, Chapter 4.
9. Pluralism and Corporatism will be referred to again in the Interest Group Chapter. The definition for the terms is drawn from Schmitter (1979).
10. Noted in several interviews and newspaper articles.

## 2 Risk—A Contested Concept

1. 'Prime Minister launches major campaign to get 'UK online' issued on 11 September 2000 and obtained at www.number-10.gov.uk/output/Page2851.asp.
2. Hutter in her study of British Rail, for instance, noted three levels of corporate response to risk regulation that charts degrees of integration and normalization of risk processes and practices (2001, 302).
3. See C. Bentley (2002, 83–93) for an outline of risk management in PRINCE2.

4. There are many definitions of the development life cycle. The National Institute of Standards in Technology (NIST) defines it for instance as Initiation—Development/Acquisition—Implementation—Operation/Maintenance—Disposal (NIST Contingency Planning, 2002, 11).
5. See CIO Council website, cio.gov.
6. Note that CBAs are often contrasted with the Precautionary Principle (Sunstein, 2005, 2007), which is treated in the next section. CBAs can be thought of as group decision-making tools; however, the collective decision tends to be the sum of the individual perspectives. It relies heavily on market signals to determine value. The Precautionary Principle, on the other hand, is almost exclusively associated with collective decision-making and an inability to arrive at a collective articulation of costs and/or benefits.
7. Douglas sees the micro and macro approach equally.
8. I write that this side of the figure is more 'optimistic' because change is possible—the problems are objective and can be understood and agreed upon. That noted, Habermas was not noted as believing that the power structures in society, which serve an elite and discriminate and oppress most, would change easily if at all.

## 3   How Did the Governments React to Y2K?

1. The cybernetic view of control (that is, information-gathering, standard-setting and behaviour modification) will be referred to throughout each of the sections.
2. In the United States, the approach was largely the same as the UK but the terms often employed in publications were: awareness; assessment; renovation; validation; and implementation.
3. Renamed the Government Accountability Office.
4. For the UK, see, for example, CCTA, 1997a,b,c,d,c,e,f—For the United States, see GAO, 1997a,b,c,d,e,f,g.
5. The UK government prioritized the sectors in terms of their dependencies on each other. The sectors with the highest level of dependency on them were placed in Tranche 1. Tranche 1 had to report its Y2K status first, with Tranche 2 to follow a few weeks later and so on. This classification will be discussed further in the Interests chapter.
6. Micro-businesses are defined as organizations with fewer than ten employees; SMEs have 10–250 employees.
7. In both countries more than 80 per cent of the critical infrastructure is owned and operated by the private sector.
8. According to interview subjects in both the United States and the United Kingdom this level of cooperation is very unusual (INT 39; 60).
9. One interview subject who attended the congressional hearing noted it was a gruelling session and had a significant impact on the Administrator.
10. Defined by passengers and cargo carried.
11. The President appoints and Congress approves Inspectors General for each government department. The *Inspector General Act* (1978) gives the Office of the Inspector General (OIG) autonomy to audit the department (and related agencies) for which the IG is responsible without interference.

12. The five categories are not mutually exclusive.
13. The FAA announced it was compliant on 21 July 1999 (DOT, 1999). The CAA announced it was compliant on 12 July 1999 (CAA, 1999).

## 4  The Market Failure Hypothesis

1. Although the figures may seem slightly dated (1994 and 1995, respectively), they were among the most up-to-date figures that the World Bank and OECD had in 1998, and were used by these organizations in publications about Y2K.
2. See, for example, NAO 1997, 1998, 1999a,b; House of Commons Library 1998; Parliamentary Office on Science and Technology (POST) 1996 and 1997; Couffou, 1997; United States Senate Special Committee on the Year 2000 Technology Problem, 1999a,b; 2000.
3. For examples, see Blitz *et al.* (1999) on Italy, Spain and Greece, Jack (1999) on Russia, Bowen (1999) on Latin America and Suzman (1999) on Eastern Europe and Indonesia.
4. For a discussion of local government readiness, see Shillingford (1999).
5. For a discussion of SMEs, see NAO (1999b, 24).
6. Y2K Advertising brochure in agency Y2K file.
7. See, for example, Hammam, 1998; *New York Times* Editorial Desk, 1999a; United States Senate Committee on Year 2000 Technology Problem, 1999a, 155.
8. Independent report conducted by solicitors Cameron McKenna for ABI (Adams, FT, 21 July 1998).
9. Council Directive 85/374/EEC of 25 July 1985 on the approximation of the laws, regulations and administrative provisions of the Member States concerning liability for defective products OJ 1985 L 210, p. 29, implemented in the United Kingdom by the Consumer Protection Act 1987.
10. 'Year 2000 Presents IT Organizations with Challenges and Opportunities', *Sun Journal*.
11. In other fields, notably the environment, the precautionary principle seems to be more popularly accepted in continental Europe and particularly in Germany. At the time it seemed to be rising in acceptance in the United Kingdom, given events such as BSE. But it was less commonly adopted in the United States.
12. The CAA and FAA did do some PRAs as matter of their regular reporting requirements but it is not clear that these forms had any impact on risk management decisions.

## 5  Opinion-Responsive Hypothesis

1. Lee Clarke notes in Worst Cases (2005) that governments often assume that people will overreact and even go hysterical during a crisis. Yet, evidence suggests the contrary; in most case people tend to be helpful and supportive to one another during crises.
2. By popular opinion polls, I mean surveys that attempted to determine the opinions of the population as a whole. Many surveys, for instance, focused primarily on the opinions of executives or IT directors only.

3. The method used for identifying articles in *Computerworld* is slightly different than that used for the other sources. For the *Computerworld* search, the key search terms could appear in the title of the article or the abstract, whereas for the others, articles were identified by key words appearing in the title only. See Methods section for additional notes.

4. This analysis is meant to give a flavour of the coverage on Y2K in the selected sources. *Government Computing* is a monthly magazine and much shorter than *Computerworld*, which is a weekly publication, so direct comparison in volume of coverage between the two would be misleading.

5. Sun articles included, 'Y2K Bug was Mouse' (6 January 2000) (2000a) and 'Millennium Bug? Well some Bloke's Video Broke Down in Pittenweem' (6 January 2000) (2000b).

6. Includes the *Sunday Times*.

7. See, for example, GMIT Subcommittee, 1999, which summarizes letter grades issued to departments between 1996 and 1999; and Science and Technology Committee, 1997/1998.

8. See Tversky, A. and Kahneman, D. (1973). 'Availability: A Heuristic for Judging Frequency and Probability'. *Cognitive Psychology* 5, 207–32.

9. Some sources in the articles may have worked for academic institutions; however, their academic affiliations were not disclosed.

10. It was a convention to programme '9999' to refer to the end of a programming loop. IT specialists speculated that the '9999' programming line would raise problems on 9 September 1999 (9999).

11. For examples, see Blitz *et al.* (1999) on Italy, Spain and Greece, Jack (1999) on Russia, Wilson (1999) on Latin America and Suzman (1999) on Eastern Europe and Indonesia.

12. Hyatt, M. (1998) *The Millennium Bug: How to Survive the Coming Chaos.* Nashville: Thomas Nelson Inc. Thomas Nelson is the biggest publisher of religious books in the United States. It should be noted that some Evangelical Christian groups, particularly in the United States, associated Y2K and the possibility of large-scale chaos as a sign of Millenarianism (Miller, 1999). Millenarianism is 'the belief in a future millennium during which Christ will reign on earth, based on Revelation 20:1–5' (taken from Collins Concise English Dictionary, 1993).

13. Both governments' strategies echo Schattschneider's notion of mobilization bias (1960). Hood *et al.* (2001), however, do not refer to Schattschneider. They do cite Dunleavy's (1991) Preference-Shaping Model.

# 6   The Interest Group Hypothesis (The Issue Network)

1. Devolution in the United Kingdom occurred in the late stages of Y2K planning. While the devolved authorities were responsible for their own Y2K operations, by and large, they accepted the plans they inherited from the central government.

2. There was a degree of continuity in Congress in 1997 by virtue of the fact that members of Congress participated in several committees, and therefore,

they and their staff developed an awareness of the far-reaching nature of the issue.

3. I analysed the statements only and not the questions and answers that followed.

4. IATA, the industry association, also started playing a large role both in the United States and abroad in gathering information on compliance at this time.

5. Not necessarily the agency case studies.

6. Recall one organization dropped out of the UK/NIF in the aviation sector yet the government still reported the sector as being Y2K compliant (Action, 2000, 1999; NAO, 1999b).

7. The FAA was not obligated to appoint a CIO under Clinger-Cohen because it is an agency, not a department. Clinger-Cohen affects departments only.

8. It might be argued that anxiety was at its peak because most organizations were behind in the assigned task, media coverage was intense and mostly negative and qualified human resources were perceived as scarce. One year later most of the anxiety had subsided.

9. Some of the organizations were manufacturers based overseas. The CAA concluded that *if* Y2K turned out to be problem with similar manufacturers then it would act in January 2000 (INT 22).

10. Some IT interview subjects noted it came too late in the process. By the time the Y2K Acts were enacted they had already spent hours in meetings with lawyers developing corporate plans.

11. The interview was not formally documented.

12. This funding strategy was noted in several interviews in the United States although no one interview subject felt this problem ever became excessive.

13. Ironically, there were also examples of departments that retained equipment that demonstrated vulnerability because staff happened to like the equipment.

14. Note that I promised interview subjects from the IT industry that I would protect their identities. Therefore information cited from their interviews will not be attributed to an interview lest it imply who might have revealed the information. Otherwise these interviews followed exactly the same process as all other interviews. For further information please see the Methods chapter.

15. See website itaa.org.

16. Promotional material obtained from Y2K agency file.

17. This list of IT service providers was obtained at Dunleavy *et al.* (2001, 11). Dunleavy *et al.* also note that IT service provision is more heavily concentrated in the UK government than in the US government. The United Kingdom tends to contract with a relatively small group of large private IT service providers whereas the US government depends on a much larger number in a variety of arrangements.

18. I include companies' names to serve as examples. In fact, large IT companies often provide numerous services. IBM, for instance, provides hardware, software and services.

19. See, for example, Kedrosky (1998) and Jones (1998).

20. In one instance, the GAO (1999c) was highly critical of the FAA for off-shoring work on classified computer code for Y2K fixes.

## 7   Conclusion

1. In 1976 in response to an anticipated outbreak of swine flu the US government orchestrated a large and far-reaching vaccination programme. The programme was enacted quickly and absorbed considerable resources. Some people died as a consequence of the exposure to the vaccine. Ultimately, however, swine flu never materialized. As well as Neustadt and Fineberg's (1983) account, it is also used as a public management case study in Moore's (1995) *Creating Public Value*.
2. Perrow still advocates CBAs to help moderate responses.
3. Many at the FAA recall their Y2K work as the highlight of their careers.
4. Rational Actor; Organizational Behaviour and Government Politics.
5. This disjunction between IT and reporting mechanisms in government is similar to observations by Bellamy and Taylor (1998, 146–70).
6. It is not strictly the case that the government *had* to do anything. Their decision to act may reflect 'anticipationism', which suggests a predisposition to acting in the face of uncertainties (Hood, 1996).
7. See, for example, James, 1999.
8. This dynamic has been noted by others, for example Rochlin (1997).
9. This is consistent with Vogel (1986). Vogel examined government/industry relations in environmental policy.

## Appendix A   Methodology

1. In its 1998 report, the OECD, citing research from Gartner, described the following countries as being the most advanced in Y2K preparations: Group 1 (most advanced)—US, Australia and Canada; Group 2 (three months behind Group 1): Holland, Belgium and Sweden; Group 3 (six months behind Group 1): UK, South Africa, Israel and Ireland (OECD, 1998, 17).

# Bibliography

## Notes

1. For some publications no author was identified. In those cases the publications have been alphabetized by the publications' names.
2. Audits that were conducted by private sector auditors are not documented in the bibliography.

Action 2000 (1999), *National Infrastructure Protection Programme: Final Report*. London: Action 2000.

Adams, C. (1997a), 'US Supplier Sued Over Millennium Bomb Till—Retailers Suit a Threat to Insurance Company'. *Financial Times*. 18 August.

Adams, C. (1997b), 'Insurers Act to Stop Claims Over Millennium Bomb'. *Financial Times*. 20 August.

Adams, C. (1997c), 'Insurers Agree Millennium Bomb Exclusion Clauses'. *Financial Times*. 10 November.

Adams, C. (1997d), 'US Insurers Limit Millennium Bomb Losses'. *Financial Times*. 22 December.

Adams, C. (1998), 'Insurers Rush to Cut the Risk Posed by Millennium Bomb'. *Financial Times*. 21 July.

Adams, J. (1995), *Risk*. London: UCL.

Allison, G. and Zelikow, P. (1999), *The Essence of Decision: Explaining the Cuban Missile Crisis*. Second Edition. Harlow: Longman.

Associated Press (1999), 'GTE Sues Five Insurers to Recover Expenses Related to Y2K'. *Wall Street Journal*. 6 July.

Atwood, L. and Major, A. (2000), 'Optimism, Pessimism, and Communication Behaviour in Response to an Earthquake Prediction'. *Public Understanding of Science*, 9: 17–431.

Aucoin, P. (1997), 'The Design of Public Organizations for the 21st c: Why Bureaucracy will Surive in Public Management'. *Canadian Public Administration*. 40,2: 290–306.

Auerbach, J. (1999), 'Y2K Problem: Y2K-Bug Fixers See Work Dive as 2000 Nears'. *Wall Street Journal*. 11 March.

Auerswald, P. E., La Porte, T. M. and Michel-Kerjan, E. O. (2006), *Seeds of Disaster, Roots of Response: How Private Action Can Reduce Public Vulnerability*. Cambridge: Cambridge University Press.

Aviram, A. (2005), 'Network responses to network threats: the evolution into private cybersecurity associations' in M. F. Grady and F. Parisi (eds), *The Law and Economics of Cybersecurity*. New York: Cambridge University Press, 143–92.

Aviram, A. and Tor, A. (2004), 'Overcoming Impediments to Information Sharing'. *Alabama Law Review*, 55: 231.

Beck, U. *(1992), Risk Society. London: Sage.*

Beckett, M. (2000a), 'Things Did Not Go Right by Accident, Bug-buster Beckett Reveals'. News Release on 20 January. London: Cabinet Office Press Office.

Beckett, M. (2000b), 'Beckett Says the Challenge Now is to Exploit the Bug Benefits'. News Release on 18 April. London: Cabinet Office Press Office.

Bellamy, C. and Taylor, J. (1998), *Governing in the Information Age*. Buckingham: Open University Press.

Bentley, C. (2002), *Practical PRINCE2*. London: TSO.

Blair, T. (1998), 'Speech by the Right Honourable Tony Blair MP on The Millennium Bug to the Action 2000/Midland Bank Conference at the Barbican on Monday 30 March 1998'. 10 Downing Street Press Notice obtained at the House of Commons Library.

Blitz, J., Hope, K. and White, D. (1999), 'Mediterranean Trio See Millennium Bug as Problem for Manana'. *Financial Times*. 12 November.

Boin, A. and McConnell A. (2007), 'Preparing for Critical Infrastructure Breakdowns: The Limits of Crisis Management and the Need for Resilience'. *Journal of Contingencies and Crisis Management*, 15: 1, 50–9.

Bowen *et al.* (1999), 'Low Tech Culture May Prove Region's Y2K Saviour'. *Financial Times*. 30 November.

Bray, R. (1999), 'Guru Gets Airborne to Bury Millennium Fears'. *Financial Times*. July 7.

Bulkeley, W. and Hamilton, D. (1999), 'The Real Y2K Problem-Computer Sales'. *Wall Street Journal*. 22 October.

Burgess, A. (2004), *Cellular Phones, Public Fears and, and a Culture of Precaution*. Cambridge: Cambridge University Press.

Cabinet Office (2000), *Successful IT: Modernising Government in Action*. London: HMSO.

Cabinet Office (2002), *Risk: Improving Government's Capability to Handle Risk and Uncertainty*. London: HMSO.

Cabinet Office (2007), 'Introduction to the Civil Contingencies Secretariat'. UK Government. http://www.ukresilience.info/ccs/aims.aspx.

Caffrey, A. (1999), 'Heard in New England: EMC Sales May Feel Bit of a Bite on Y2K Fears, Analysts Predict'. *Wall Street Journal*. 2 February.

Cane, A. (1998), 'Minister May Seek Recovery of Millennium Bomb Costs'. *Financial Times*. 16 June.

Cartiglia, F., Patelis, A. and Perez de Azpillage, J. (1998), *Is the Millennium Bug a Danger? Y2K: A Tremor, not a Quake* Goldman Sachs. Obtained at the British House of Commons Research Room. Filed under Basic Science and Technology Indicators, S100–79.

Central Computer and Telecommunications Agency (1997a), *Managing the Programme: Reach Out I'll Be There*. Cambridge: Cambridge Publishers.

Central Computer and Telecommunications Agency (1997b), *Kick-Starting the Organization*. Cambridge: Cambridge Publishers.

Central Computer and Telecommunications Agency (1997c), *Assessing the Size of Your Problem: River Deep, Mountain High*. Cambridge: Cambridge Publishers.

Central Computer and Telecommunications Agency (1997d), *Tackling the Year 2000: The Legal Implications*. Cambridge: Cambridge Publishers.

Central Computer and Telecommunications Agency (1997e), *Testing and Compliance: The Final Countdown*. Cambridge: Cambridge Publishers.

Central Computer and Telecommunications Agency (1997f), *An Executive Overview: Time is Tight*. Cambridge: Cambridge Publishers.

Chief Information Officer Council www.cio.gov.

Civil Aviation Authority (CAA) (1998), 'Year 2000'. *Memo to Aviation Industry.* 8 July.

Civil Aviation Authority (CAA) (1999), 'UK Aviation Industry Now Ready for Year 2000'. *CAA News Release,* 12 July.

Clarke, L. (2005), *Worst Cases: Terror and Catastrophe in the Popular Imagination.* Chicago: University of Chicago Press.

Clinton, W. (1998), 'Remarks by the President Concerning the Year 2000 Conversion'. Address at the National Academy of Sciences. 14 July. www.techlawjournal.com.

Cm 1730 (1991), *Competing for Quality.* London: HMSO.

Cm 3438 (1996), *Government. Direct.* London: HMSO.

Cm 4310 (1999), *Modernising Government.* London: HMSO.

Cm 4703 (2000), *Modernising Government in Action: Realising the Benefits of Y2K.* London: HMSO.

Cohen, S. (1972), *Folk Devils and Moral Panics.* Oxford: Martin Robertson.

Collier's *Encyclopedia.* (1983). New York: Collier's.

*Collins Concise English Dictionary* (1993), J. Sinclair (ed.) Wrotham: Harper Collins.

Committee on Public Accounts (PAC) (1999), *The Millennium Threat,* 36th Report. London.

Comfort, L. (2002), 'Rethinking Security: Organizational Fragility in Extreme Events'. *Public Administration Review,* 62: 1, 98–107.

Couffou, A. (1997), 'Year 2000 risks: what are the consequences of technology failure?' *Statement of Hearing Testimony before Subcommittee on Technology and Subcommittee on Government Management, Information and Technology.* 20 March. Washington, DC: GPO.

Dake, K. (1991), 'Orienting Dispositions in the Perception of Risk: An Analysis of Contemporary Worldviews and Cultural Biases'. *Journal of Cross-Cultural Psychology,* 22: 61–82.

de Bruijne, M. and van Eeten, M. (2007), 'Systems that Should Have Failed: Critical Infrastructure Protection in an Institutionally Fragmented Environment'. *Journal of Contingencies and Crisis Management,* 15: 1, 18–29.

de Jager, P. (1993), 'Doomsday 2000'. *Computerworld.* 16 September.

de Jager, P. www.technobility.com.

Delaney, K. (1999), 'Y2K Tests, Upgrades Cause their Own Chaos'. *Wall Street Journal.* 7 September.

Department of Trade and Industry (DTI) (1996), *Ian Taylor Welcomes Date Change 2000 Task Force.* DTI Press Notice Obtained from the House of Commons Library.

Department of Transportation (DOT) (1999), 'FAA is 100 per cent Y2K Compliant'. *DOT Newsrelease.* 21 July.

Dietz, T. and Stern, P. (1995), 'Towards a Theory of Choice: Socially Embedded Preferences Construction'. *Journal of Socio-Economics,* 24: 261–79.

Douglas, M. (1982), *In the Active Voice.* London: Routledge.

Douglas, M. (1992), *Risk and Blame: Essays in Cultural Theory.* London: Routledge.

Dowding, K. (1995), 'Model or Metaphor? A Critical Review of the Policy Network Approach'. *Political Studies,* 43: 136,158.

Downs, A. (1972), 'Up and Down with Ecology: The Issue Attention Cycle'. *Public Interest,* 28:1, 38–50.

Dunleavy, P. (1991), *Democracy, Bureaucracy and Public Choice*. Essex: Prentice Hall.
Dunleavy, P., Margetts, H., Bastow, S., Tinkler, J. and Yared, H. (2001), 'Policy Learning and Public Sector Information Technology'. Paper for the American Political Science Association's Annual Conference 2001, San Francisco.
Efficiency Unit (1988) (Ibbs Report), *Improving Management in Government: The Next Steps*. London: HMSO.
Egan, M. J. (2007), 'Anticipating Future Vulnerability: Defining Characteristics of Increasingly Critical Infrastructure-like Systems'. *Journal of Contingencies and Crisis Management*, 15: 1, 4–17.
Ernst and Young (1998), *Millennium Infrastructure Project*. Ernst and Young for Cabinet Office and Obtained at the House of Commons Library.
Ezell, B. C. (2007), 'Infrastructure Vulnerability Assessment Model'. *Risk Analysis*, 27: 3, 571–84.
*Farmer's Almanac* (2000), Almanac Publishing. www.farmersalmanac.com.
Feder, B. (1999a), 'The Dominant Position of the Gartner Group'. *New York Times*. 5 July.
Feder, B. (1999b), 'Keeping Home PCs Afloat Through Y2K'. *New York Times*. 28 October.
Feder, B. (1999c), 'For Worriers, Winding Down on Year 2000'. *New York Times*. 27 December.
*Financial Times* (1998), 'KLM Says Millennium Bug May Ground its Flights'. 17 October.
Finkelstein, A. (2000), 'Y2K: A Retrospective View'. www.cs.ucl.ac.uk/Staff/ A.Finkelstein, originally published in *Computing and Control Engineering Journal*, August 2000, v 11, N4, 156–9.
Fischhoff, B. (1985), 'Managing Risk Perception'. *Issues in Science and Technology*. 2: 3–96.
Fischhoff, B. (1995), 'Risk Perception and Communication Unplugged: Twenty Years of Process'. *Risk Analysis*, 15: 137–45.
Fountain, J. (1999), 'Paradoxes of Public Sector Customer Service'. www.harvard. edu in May 2001.
Fountain, J. (2001), *Building the Virtual State: Information Technology and Institutional Change*. Washington, DC: Brookings.
Freudenberg, W.R., Coleman, C.L., Gonzales, J. and Helgeland, C. (1996), 'Media Coverage of Hazard Events: Analysing the Assumptions'. *Risk Analysis*, 16: 31–42.
Gallup Organization (1999), Y2K Polls. Gallup. www.gallup.com.
Gantz, J. (1997), 'How Big is the Gorilla on your Desk?' *Computerworld*, 31 March. 31: 13.
Gaskell, G. Bauer, M. Durant, J. and Allum, N. (1999), 'Worlds Apart? The Reception of Genetically Modified Foods in Europe and the US'. *Science*, 285:16. July, 384–7.
General Accounting Office (1997a), *Year 2000 Computing Crisis: Strong Leadership Today Needed to Prevent Future Disruption of Government Services*. GAO/T-AIMD-97-51, 24 February. Washington, DC: GAO.
General Accounting Office (1997b), *Year 2000 Computing Crisis: Risk of Serious Disruption to Essential Government Functions Calls for Agency Action Now*. GAO/ T-AIMD-97-52, 27 February. Washington, DC: GAO.
General Accounting Office (1997c), *High-Risk Series: Information Management and Technology*. GAO/HR-97-9, February. Washington, DC: GAO.

General Accounting Office (1997d), *Year 2000 Computing Crisis: Time is Running Out for Federal Agencies to Prepare for the New Millennium.* GAO/T-AIMD-97-129, 10 July. Washington, DC: GAO.

General Accounting Office (1997e), *Year 2000 Computing Crisis: Success Depends upon Strong Management and Structured Approach.* GAO/T-AIMD-97-173, 25 September. Washington, DC: GAO.

General Accounting Office (1997f), *Year 2000 Computing Crisis: An Assessment Guide.* GAO/AIMD-10.1.14, September. Washington, DC: GAO.

General Accounting Office (1997g), *Year 2000 Computing Crisis: Risk of Serious Disruption to Essential Government Functions Calls for Action Now.* GAO:/AIMD-97-52, 27 February. Washington, DC: GAO.

General Accounting Office (1998a), *Year 2000 Computing Crisis: Status of Airports Efforts to Deal with Date-Change Problem.* GAO/RCED/AIMD-99-57. 29 January. Washington, DC: GAO.

General Accounting Office (1998b), *FAA Computer Systems: Limited Progress on Year 2000 Issue Increases Risk Dramatically.* GAO/AIMD-98-45. 30 January. Washington, DC: GAO.

General Accounting Office (1998c), *FAA Systems: Serious Challenges Remain in Resolving Year 2000 and Computer Security Problems.* GAO/T-AIMD-98-251. 6 August. Washington, DC: GAO.

General Accounting Office (1998d), *Year 2000 Computing Crisis: Progress Made at Labor, but Key Systems at Risk.* GAO/T-AIMD-98-303. 17 September. Washington, DC: GAO.

General Accounting Office (1999a), *Year 2000 Computing Challenge: Readiness Improving, but Critical Risks Remain.* GAO/T-AIMD-99-49. 20 January. Washington, DC: GAO.

General Accounting Office (1999b), *Year 2000 Computing Challenge: Labor Has Progressed but Selected Systems Remain at Risk.* GAO/T-AIMD-99-179. 12 May. Washington, DC: GAO.

General Accounting Office (1999c), *Computer Security: FAA Needs to Improve Controls over Use of Foreign Nationals to Remediate and Review Software.* GAO:/AIMD-00-55, 23 December. Washington, DC: GAO.

General Accounting Office (2000), *Year 2000 Computing Challenge: Lessons Learned can be Applied to Other Management Challenges.* GAO:/AIMD-00-290. September. Washington, DC: GAO.

Gomes, L. (1999), 'New Year's Revels in Y2K War Room; Fear of Flying? Air Carriers Ground Flights on 12/31, Cut by 20%'. *Wall Street Journal.* 2 December.

Government Accountability Office (2001), *Information-Sharing: Practices that Can Benefit Critical Infrastructure Protection.* GAO-02-24. Washington, DC: GAO.

Government Accountability Office (2003), *Critical Infrastructure Protection: Challenges for Selected Agencies and Industry Sectors.* GAO-03-233. Washington, DC: GAO.

Government Accountability Office (2004a), *Critical Infrastructure Protection: Improving Information Sharing with Infrastructure Sectors.* GAO-04-780. Washington, DC: GAO.

General Accounting Office (2004b), *Leadership Remains Key to Agencies Making Progress on Enterprise Architecture Efforts.* Washington, DC: GAO.

General Accounting Office (2004c), *Federal Agencies Have Made Progress Implementing the E-Government Act of 2002.* Washington, DC: GAO.

Gore, A. (1993), *From Red Tape to Results*. Washington, DC: GAO.

Government Management, Information and Technology Subcommittee (GMIT) (1999), 'Year 2000 Progress Report Card'. www.house.gov/refor/amit.

Graham, G. (1997), 'RBS Targets the Millennium Bug—Bank Taking a £29 M Charge Despite Advice Against Move'. *Financial Times*. 28 November.

Graham, J. (2002), 'The Role of Precaution in Risk Management'. Remarks Prepared For the International Society of Regulatory Toxicology and Pharmacology Precautionary Principle Workshop in Crystal City, VA. 20 June.

Grande, C. (2000), 'Lessons for Business in Measures Against the Millennium Bomb'. *Financial Times*. 1 March.

Gutteling, J. and Kuttschreuter, M (2002), 'The Role of Expertise in Risk Communication: Laypeople's and Expert's Perception of the Millennium Bug Risk in the Netherlands'. *Journal of Risk Research*, 51: 35–47.

Habermas, J. (1984), *Reason and the Rationalization of Society*. Boston: Beacon Press.

Hall, B. (1997), Testimony to Government Management Information and Technology Subcommittee. 20 March.

Hammam (1998), '2000 Blessings on Lawyer—Millennium Bug May Not be a Disaster for Everyone'. *Financial Times*. 20 February.

Hansard (1999), Column 184. 25 November. London: HMSO.

Haught, D. (1998), *Solving the Year 2000 Problem in Microsoft Desktop Application Programs*. FMS Technical Papers. www.fmsinc.com.

Health and Safety Executive (1998), *Risk Assessment and Risk Management*. (Second Report Prepared by the Interdepartmental Liaison Group on Risk Assessment). London: Health and Safety Executive.

Heclo, H. (1978), 'Issue Networks and the Executive Establishment' in A. King (ed.) *The New American Political System*. Washington, DC: American Enterprise Institute.

Heeks, R. (1999), *Reinventing Government in the Information Age*. London: Routledge.

HM Treasury (2004), *The Orange Book: Management of Risk–Principles and Concepts*. London: HMSO.

HM Treasury (2005), *Managing Risks to the Public*. London: HMSO.

HM Treasury and Office of Government Commerce (OGC) (2005), *Managing Risks with Delivery Partners*. London: HMSO.

Hodson, M. (1998), 'Destination: Disaster'. *London Times* (Travel). 22 November.

Hoffman, T. (1999), 'Y2K Failures have Hit 75% of US Firms'. *Computerworld*. 16 August, 33: 33.

Hood, C. (1991), 'A Public Management for All Seasons?'. *Public Administration*, 69: 1, 3–19.

Hood, C. (1996), ' "Where extremes meet: SPRAT versus SHARK" in public risk management' in C. Hood and D. Jones (eds), *Accident and Design*. London: UCL.

Hood, C. (1998), *The Art of the State*. Oxford: Clarendon.

Hood, C., Rothstein, H. and Baldwin, R. (2001), *The Government of Risk: Understanding Risk Regulation Regimes*. Oxford: Oxford University Press.

House of Commons Library (1998), *The Millennium Bug*, Research Paper: 98/72. London: House of Commons.

Howells, G. (1999), 'The Millennium Bug and Product Liability'. *Journal of Information, Law and Technology*. elj.warwick.ac.uk/jilt/99-2.

Hutter, B. (2001), *Regulation and Risk: Occupational Health and Safety on the Railways*. Oxford: Oxford University Press.

International Air Transport Association (IATA). www.iata.org. International Data Corporation (1998), found at: file:///Users/eduni/Documents/ Dissertation/ MFH%20Chapter/Y2K%20Cost%20Estimates.html.

Jack, A. (1999), 'Level of Russian IT Lessens Year 2000 Fears: Moscow Has Been Late in Putting Together Some Moderate Y2K Defences'. *Financial Times*. 26 November.

Jacobs, M. (1998), 'Big Companies Swear off Lawsuits to Minimise Cost of Y2K Problem'. *Wall Street Journal*. 30 November.

Jaeger, C., Renn, O., Rosa, E. and Webler, T. (2001), *Risk, Uncertainty and Rational Action*. London: Earthscan.

James, S. (1999), *British Cabinet Government*. London: Routledge.

Johnson, B. B. and Cavello, V. T. (1987), *The Social and Cultural Construction of Risk*. Dordrecht: Reidel.

Jones, A. (1998), 'Millennium Bug Pays Off for Select Few'. *London Times*. 6 November.

Jones, D. (1999), 'Y2K: From Horror to Ho-Hum'. *USA Today*. 22 December.

Kasperson, R. (1992), 'The Social Amplification of Risk: Progress in Developing an Integrative Framework' in S. Krimsky and D. Golding (eds), *Social Theories of Risk* London: Praeger.

Kedrosky, P. (1998), 'To Figure Out Y2K Hype, Follow the Money'. *Wall Street Journal*. 20 July.

Kehoe, L. (1999), 'Millennium Bug Hits IBM Shares'. *Financial Times*. 22 October.

Kelly, J. (1998), 'Lloyds Launches Millennium Bomb Cover'. *Financial Times*. 2 September.

Kheifets, L., Hester, G. and Banerjee, G. (2001), 'The Precautionary Principle and EMF: Implementation and Evaluation'. *Journal of Risk Research*. 4(2): 113–125.

King, J. (1999), 'EDS Y2K About-Face Raises WIN 95 Doubts'. *Computerworld*. 33: 13.

Kitzinger, J. (1999), 'Researching Risk and the Media'. *Health, Risk and Society*. 1: 1, 55–69.

Kitzinger J. and Reilly, J. (1997), 'The Rise and Fall of Risk Reporting: Media Coverage of Human Genetics Research, False Memory Syndrome and Mad Cow Disease'. *European Journal of Communication*. 12: 319–50.

Lagnado, L. Rhundle, R., Carrns, A. and Conklin, J. C. (1999), 'Y2K Fixing the Bug: Hospitals Take No Chances in Planning'. *Wall Street Journal*. 27 December.

LaPorte, T. (1996), 'High Reliability Organizations: Unlikely, Demanding and At Risk'. *Journal of Crisis and Contingency Management*, 4: 2, 60–71.

LaPorte, T. R. and Consolini, P. (1991), 'Working in Practice but not in Theory: Theoretical Challenges of High Reliability Organizations'. *Journal of Public Administration Research and Theory*, 1: 19–47.

Leahy, P. and Mazur, A. (1980), 'The Rise and Fall of Public Opposition in Specific Social Movements'. *Social Studies of Science*, 10: 259–84.

Leake, J. (1998), 'Air Chaos Feared as New Computer Fails'. *Times*. 5 April.

Lichtblau, E. (2005), 'Security Report on US Aviation Warns of Holes'. *New York Times*. 14 March.

Lohse, D. (1999a), 'Insurers Y2K Payout is Pegged Above $15 B'. *Wall Street Journal*. 21 June.

Lohse, D. (1999b), 'Insurance Industry and Corporations Battle Over Reimbursement for Y2K Bug'. *Wall Street Journal*. 16 August.

Manion, M. and Evans, W. (2000), 'The Y2K Problem and Professional Responsibility: A Retrospective Analysis'. *Technology in Society*. 22: 3, 361–87.

Lowrance, W. (1976), *Of Acceptable Risk: Science and the Determination of Safety*. Los Altos: William Kaufman.

Luhmann, N. (1993), *Risk: A Sociological Theory*. New York: Aldine de Gruyter.

Margetts, H. (1999), *Information Technology in Government: Britain and America*. London: Routledge.

Marsh, D. and Rhodes, R.A.W. (1992), *Policy Networks in British Government*. Oxford: Oxford University Press.

McCarthy, J. A. (2007), 'From Protection to Resilience: Injecting Moxie into the Infrastructure Security Continuum'. *Critical Thinking: Moving from Infrastructure Protection to Infrastructure Resilience*. CIPP Discussion Paper Series. George Mason University. Critical Infrastructure Protection Program. http://cipp.gmu.edu/

McGough, R. (1999), 'As Y2K Stock Plays Flop, Investors Seek New Angles'. *Wall Street Journal*. 17 September.

Merchant, K. (2000), 'Indian Software Suppliers Build on Y2K Success'. *The Financial Times*.

Micklethwait, J. and Woolridge, A. (1996), *The Witch Doctors: Making Sense of the Management Gurus*. New York: Random House.

Middleton, O. (1999), 'FDA Says Y2K Bug Could Affect Few Medical Devices'. *Wall Street Journal*. 27 August.

Miller, L. (1999), 'Apocalypse No'. *Wall Street Journal*. 21 January.

Moore, M. (1995), *Creating Public Value: Strategic Management in Government*. Cambridge: Harvard University Press.

Moore, M. (2003), *Bowling for Columbine*. Salter Street Films.

MORI (1999), *Y2K Polls*. www.mori.com.

Moynihan, D. P. (1996), 'Y2K Letter to President Clinton'. Entered in the Congressional Record 11 August.

Murdock, G., Petts, J. and Horlick-Jones, T. (2003), 'After amplification: rethinking the role of the media in risk communication' in N. Pidgeon, R. Kasprson, P. Slovic (eds), *The Social Amplification of Risk*. Cambridge: Cambridge University Press. 156–78.

Murray, S. K. and Howard, P. (2002), 'Variation in White House Polling Operations'. *Public Opinion Quarterly*. 66: 527–58.

Mutz, D. and Soss, J. (1997), 'Reading Public Opinion: The Influence of News Coverage on Perceptions of Public Sentiment'. *Public Opinion Quarterly*. 61: 431–51.

Nairn, G. (1998), 'Special Help for Those with Y2K Computer Fears'. *Financial Times*. 2 December.

National Air Traffic Services (1999a), 'NATS Declares its Year 2000 Readiness'. News Release, 29 March.

National Air Traffic Services (1999b), 'Business As Usual for NATS on Millennium Night'. News Release, 29 March.

National Audit Office (1997), *Managing the Millennium Threat*. London: HMSO.

National Audit Office (1998), *Managing the Millennium Threat II*. London: HMSO.

National Audit Office (1999a), *The Millennium Threat: 221 Days and Counting*. London: HMSO.

National Audit Office (1999b), *The Millennium Threat: Are We Ready?* London: HMSO.

National Audit Office (1999c), *United Kingdom Passport Agency: The Passport Delays of Summer 1999*. London: HMSO.

National Audit Office (1999d), *Government on the Web*. London: HMSO.

National Audit Office (2000a), *Supporting Innovation: Managing Risk in Government Departments*. London: HMSO.

National Audit Office (2000b), *The Cancellation of the Benefits Payment Card Project*. London: HMSO.

National Audit Office (2002a), *Better Public Services Through E-Government*. London: HMSO.

National Audit Office (2002b), *Government on the Web II*. London: HMSO.

National Audit Office (2003a), *Progress in Making E-Services Accessible to All: Encouraging use by Older People*. London: HMSO.

National Audit Office (2004a), *Managing Risks to Improve Public Services*. London: HMSO.

National Audit Office (2004b), *Improving IT Procurement: The Impact of Government Commerce's Initiatives on Departments and Suppliers in the Delivery of Major IT-enabled Projects*. London: HMSO.

National Institute of Standards and Technology (NIST) (2001), *Risk Management Guide for Information Technology Systems*. Special Publication 800–30. Washington, DC: GPO.

National Institute of Standards and Technology (NIST) (2002), *Contingency Planning Guide for Information Technology Systems*. Special Publication 800–34. Washington, DC: NIST.

Neustadt, R. E. and Fineberg, H. (1983), *The Epidemic That Never Was: Policy-Making and the Swine Flu Scare*. New York: Vintage Books.

New York Times Editorial Desk (1999a), 'Liability for the Millennium Bug'. *New York Times*. 26 April.

New York Times Editorial Desk (1999b), 'Liability Limits'. *New York Times*. 3 July.

OECD (1995), *Governance in Transition: Public Management Reforms in OECD Countries*. Paris: OECD.

OECD (1998), *The Year 2000: Impacts and Actions*. Paris: OECD.

Office of Government Commerce (OGC) (2001), *Draft Guidelines on Managing Risk*. London: OGC.

Office of Government Commerce (OGC) (2004), *Value for Money Measurement*. London: OGC.

Office of Management and Budget (1997a), *Getting Federal Computers Ready for 2000*. 6 February. Washington, DC: GPO.

Office of Management and Budget (1997b), *Progress on Year 2000 Conversion*. 15 August. Washington, DC: GPO.

Office of Management and Budget (1997c), *Progress on Year 2000 Conversion*. 15 November. Washington, DC: GPO.

Office of Management and Budget (1998a), *Progress on Year 2000 Conversion*. 15 May. Washington, DC: GPO.

Office of Management and Budget (1998b), *Progress on Year 2000 Conversion*. 15 August. Washington, DC: GPO.

Office of Management and Budget (1999a), *Progress on Year 2000 Conversion*. 18 March. Washington, DC: GPO.

Office of Management and Budget (1999b), *Progress on Year 2000 Conversion.* 15 June. Washington, DC: GPO.

Office of Management and Budget (1999c), *Progress on Year 2000 Conversion.* 14 December. Washington, DC: GPO.

Office of Management and Budget (2002), *E-Government Strategy.* Washington, DC: GPO.

Office of Management and Budget (2005), *Federal Information Security Management Act: 2004 Report to Congress.* 1 March. Washington, DC: GPO.

Office of National Statistics (2000), Briefing Material. Obtained at the ONS Office in Southampton.

Osborne, D. and Gaebler, T. (1992), *Reinventing Government: How the Entrepreneurial Spirit is Transforming the Public Sector.* Reading, MA: Addison-Wesley.

Page, E. (1992), *Political Authority and Bureaucratic Power: A Comparative Analysis.* London: Harvester Wheatsheaf.

Page, E. (2001), *Governing by Numbers.* Oxford: Hart Publishing.

Parliamentary Office of Science and Technology (1996), 'Computer Systems and the Millennium'. Taken from the Web in July 2001 at www.parliament.uk/post/home.

Parliamentary Office of Science and Technology (1997), 'The Millennium Threat: An Update'. Taken from the Web in July 2001 at www.parliament.uk/post/home.htm.

Perrow, C. (1999), *Normal Accidents: Living with High Technologies* (Second Edition). Princeton: Princeton University Press.

Peysner, J. (1999), 'Y2K—Will There be a Litigation Explosion?' *The Journal of Information, Law and Technology.* elj.warwick.ac.uk/jilt/99-2.

Pidd, M. (2003), *Tools for Thinking: Modelling in Management Science* (Second Edition). Chichester: Wiley & Sons.

Plain English Campaign www.plainenglish.co.uk.

Power, M. (2004), *The Risk Management of Everything.* London: Demos.

President's Council on Year 2000 Conversion (1999), *Third Summary of Assessment Information.* August. Washington, DC: GPO.

President's Council on Year 2000 Conversion (2000), *The Journey to Y2K: Final Report.* www.y2k.gov/docs/lastrep3.htm.

Quigley, K. (2002), 'The Emperor's New Computers: Y2K (Re)Visited'. *Emerging Research in Corporate Governance* (Conference Proceedings). British Accounting Association Special Interest Group on Corporate Governance, Third International Conference, Belfast: Queen's University.

Quigley, K. (2004), 'The Emperor's New Computers: Y2K (Re)Visited'. *Public Administration,* 82,4, 801–30.

Quigley, K. (2005a), 'Bug Reactions: Considering US Government and UK Government Y2K Operations in Light of Media Coverage and Public Opinion Polls'. *Health, Risk and Society.* 7.3, 267–291.

Quigley, K. (2005b), 'Risk regulation regimes in aviation: were the chips ever really down in the UK's management of Y2K?' in I. Demirag (ed.) *Towards Better Regulation, Governance and Accountability: Global Perspectives from Corporations and Civil Society.* Sheffield: Greenleaf Publishing.

Quigley, K. (2006), 'Critical Infrastructure Protection in Comparative Perspective'. *A Performing Public Sector: The Second Transatlantic Dialogue.* Hosted Jointly by the European Group on Public Administration and the American Society for Public Administration at the Katholieke Universiteit Leuven, Belguim. June.

Paper available at: http://soc.kuleuven.be/io/performance/program/workshops/ workshop3.htm.

Quigley, K. (2008), 'Planning for the worst, bringing out the best? Lessons from Y2K'. *Social Information Technology: Connecting Society and Cultural Issues.* Hershey, PA: Idea Group Publishing.

Railway Safety Standards Board (2005), *Risk Profile Bulletin.* Issue 5. London: RSSB.

Read, M. (1992), 'Policy networks and issue networks: the politics of smoking' in Marsh and Rhodes (eds), *Policy Networks in British Government.* Oxford: Clarendon Press.

Renn, O. (2008), 'Concepts of Risk: An Interdisciplinary Review'. *GAIA,* 17(1): 50–66.

Renn, O., Burns, W., Kasperson, R., Kasperson, J. and Slovic, P. (1992), 'The Social Amplification of Risk: Theoretical Foundations and Empirical Application'. *Social Issues,* 48: 137–60.

Rhodes, R.A.W. (1981), *Control and Power in Central-Local Government Relationships.* Farnborough: Gower.

Rhodes, R.A.W. (1997), *Understanding Governance: Policy Networks, Governance, Reflexivity and Accountability.* Buckingham: Open University Press.

Rochlin, G. (1997), *Trapped Inside the Net: The Unanticipated Consequences of Computerization.* Princeton: Princeton University Press.

Romano, J. (1998), 'Dealing with the Y2K Bug'. *New York Times.* 16 August.

Roux-Dufort, C. (2007), 'Is Crisis Management (Only) a Management of Exceptions?' *Journal of Contingencies and Crisis Management,* 15: 2, 105–14.

Rowland, D. (1999), 'Negligence, Professional Competence and Computer Systems'. *The Journal of Information, Law and Technology,* elj.warwick.ac.uk/ jilt/99-2.

Rowe, G., Frewer, L. and Sjoberg, L. (2000), 'Newspaper Reporting of Hazards in the UK and Sweden'. *Public Understanding of Science,* 9: 59–78.

Sagan, S. (1993), *The Limits of Safety: Organizations, Accidents, and Nuclear Weapons.* Princeton: Princeton University Press.

Schattschneider, E. (1960), *The Semisovereign People: A Realist's View of Democracy in America.* London: Holt, Rinehart and Winston.

Schlesinger, J. and McKinnon, J. (2000), 'Y2K: A Smooth Start; Demand for Cash was Normal at Banks, So Fed Will Take Back Extra Currency'. *Wall Street Journal.* 3 January.

Schmitter, P. (1979), 'Still the Century of Corporatism?'. *Trends Towards Corporatist Intermediation.* London: Sage, 7–52.

Schulman, P. R. and Roe, E. (2007), 'Designing Infrastructures: Dilemmas of Design and the Reliability of Critical Infrastructures'. *Journal of Contingencies and Crisis Management,* 15: 1, 42–9.

Sendle, E. (1998), 'Auto Makers Battle Y2K Bug in Vast Supplier Network'. *Wall Street Journal.* 30 November.

Shillingford, J. (1999), 'Some sections are Stalled at the Blue Traffic Light'. *Financial Times.* 3 November.

Simon, H. (1954), 'Some strategic considerations in the construction of social science Models' in H. Simon (ed.) (1982) *Models of Bounded Rationality: Behavioral Economics and Business Organization.* Cambridge, MA: MIT Press.

Simon, J. (1999), *British Cabinet Government.* London: Routledge.

Simons, J. (1999a), 'Trade Groups and Companies Join Forces to Push Limited Y2K Liability'. *Wall Street Journal,* 3 February.

Simons, J. (1999b), 'Senate Approves a Measure Limiting Firms' Liability Related to Y2K Glitch'. *Wall Street Journal.* 16 June.

Sjoberg, L. (1997), 'Explaining Risk Perception: An Empirical Evaluation of Cultural Theory'. *Risk Decision and Policy* 2, 2: 113–30.

Slovic, P. (1992), 'Perception of Risk: Reflections on the Psychometric Paradigm' in S. Krimsky and D. Golding (eds), *Social Theories of Risk*. London: Praeger.

Smith, A. (2000), 'Lessons for Business in Measures against the Millennium Bug'. *Financial Times.* 15 March.

Smith, R. and Buckman, R. (1999), 'Wall Street Deploys Troops to Battle Y2K: Command Centre, Jets, Extra Toilets are at the Ready'. *Wall Street Journal.* 22 December.

Soumerai, S. B., Ross-Degnan, D. and Kahn, J. S. (1992), 'Effects of Professional and Media Warnings about the Association between Aspirin use in Children and Reye's Syndrome'. *Millbank Quarterly,* 70: 155–82.

Standing Committee on Science and Technology (SCST) (1997/1998), *The Year 2000—Computer Compliance, Second Report to the House of Commons.* London: HMSO.

Starkman, D. (2000), 'Y2K Related Buying Distorts Some Companies Results; Concerns About Foul ups Lead to Stock-Piling'. *Wall Street Journal.* January.

Starr, C. (1969), 'Social Benefit Versus Technological Risk: What is Our Society Willing to Pay for Safety?' *Science,* 16: 1232–8.

*Sun (The)* (2000a), 'Y2K Bug Was Mouse'. 6 January.

*Sun (The)* (2000b), 'Millennium Bug? Well Some Bloke's Video Broke Down in Pittenweem'. 6 January.

Sunstein, C. (2005), *The Laws of Fear: Beyond the Precautionary Principle.* Cambridge: Cambridge University Press.

Sunstein, C. (2007), *Worst Case Scenarios.* Cambridge: Harvard University Press.

Suzman, M. (1999), 'World is Mostly Ready for Y2K'. *Financial Times.* 30 December.

Tan, K. (1999), 'Options Report: Volatility Index Climbs as Y2K Approaches; Traders Refrain from Making Major Moves'. *Wall Street Journal.* 31 December.

Taylor, P. (1997), 'Guide to Defuse Millennium Bomb Unveiled'. *Financial Times.* 12 May.

Taylor, P. (1998), 'Standards and SEC'. *Financial Times.* 1 September.

Taylor, P. (1999), 'Millennium Bug Fails to Materialise'. *Financial Times.* 9 October.

Taylor-Gooby, P. (2004), 'Psychology, Social Psychology and Risk'. Working Paper. Social Contexts and Responses to Risk Network, University of Kent at Canterbury. Paper obtained at http://www.kent.ac.uk/scarr/papers/papers.htm.

Thibodeau, P. (2000), 'Government Report Finds Few Y2K-Related Lawsuits'. *Computerworld,* 2 October. 34: 40.

Thompson, M., Ellis, R. and Wildavsky, A. (1990), *Cultural Theory.* Boulder CO: Westview.

United States Senate Special Committee on the Year 2000 Technology Problem (1999a), *Investigating the Impact of the Year 2000 Problem* (February Committee Report). Washington, DC: GPO. www.senate.gov.

United States Senate Special Committee on the Year 2000 Technology Problem (1999b), *Investigating the Year 2000 Problem: 100 Day Report* (September Committee Report). Washington, DC: GPO. www.senate.gov.

United States Senate Special Committee on the Year 2000 Technology Problem (2000), *Y2K Aftermath—A Crisis Averted* (Final Committee Report). Washington, DC: GPO. www.senate.gov.

US-Canada Power System Outage Task Force (2004), *Final Report*. https://reports. energy.gov/

Vaughan, D. (1996), *The Challenger Launch Decision: Risky Technology, Culture and Deviance at NASA*. Chicago: Chicago University Press.

Verba, S. and Nie, N. (1972), *Participation in America: Political Democracy and Social Equality*. Chicago: University of Chicago Press.

Vogel, D. (1986), *National Style of Regulation: Environmental Policy in Great Britain and the US*. Ithaca: Cornell University Press.

Wahlberg, A. and Sjoberg, L. (2000), 'Risk Perception and the Media'. *Journal of Risk Research*. 3,1: 31–50.

*Wall Street Journal* (1999), 'Banks Still See Low Demand for Y2K Loans'. 23 November.

Weik, K. E. (1987), 'Organizational Culture as a Source of High Reliability'. *California Management Review*. 29, 2: 112–27.

Weick, K. E. and K. M. Sutcliffe (2001), *Managing the Unexpected: Assuring High Performance in an Age of Complexity*. San Francisco: Jossey-Bass.

Wheelwright, G. (1998), 'Doomsayers of Y2K Head for the Hills'. *London Times*. 19 August.

Wilkinson, I. (2001), 'Social Theories of Risk Perception: At Once Indispensable and Insufficient'. *Current Sociology*, 49: 1–22.

Wilson, J. (1980), *The Politics of Regulation*. New York: Basic Books.

Wilson, J. (1999), 'Low Tech Culture May Prove Region's Y2K Saviour'. *Financial Times*. 30 November.

Wilson, K. (2000), 'Communicating climate change through the media: predictions, politics and perceptions of risk' in S. Allan, B. Adam and C. Carter (eds), *Environmental Risks and the Media*. London: Routledge. 201–17.

Yardeni, E. (1998), 'Y2K—An Alarmist View'. *Wall Street Journal*. 4 May.

Yardeni, E. www.yardeni.com.

Yourdon, E. www.yourdon.com.

Zinn, J. (2004), 'Literature Review: Sociology and Risk'. Working Paper. Social Contexts and Responses to Risk Network, University of Kent at Canterbury. Paper obtained at http://www.kent.ac.uk/scarr/papers/papers.htm.

Zuckerman, M. J. and Wolf, R. (1999), 'Few Worried About Y2K, But Just in Case'. *USA Today*. 24 November.

# Index

GPSR Compliance
The European Union's (EU) General Product Safety Regulation (GPSR) is a set
of rules that requires consumer products to be safe and our obligations to
ensure this.

If you have any concerns about our products, you can contact us on

ProductSafety@springernature.com

In case Publisher is established outside the EU, the EU authorized
representative is:

Springer Nature Customer Service Center GmbH
Europaplatz 3
69115 Heidelberg, Germany